Evolution Ended

The Next Stage of American Society

J.J. Jerome

Dedicated to my son Alex,

Wishing you a world where absolutely anything is possible!

Preface

We all feel it...our society seems to be going in the wrong direction. At first, I thought it was just an "old guy thing" until my 26-year son told me that he was worried about his future because "everything is going downhill." Watching the recurring stories on the evening news of mass shootings, opioid deaths, and homelessness, it seems like we have forgotten how to survive.

And yet we do survive because modern technology makes "survival of the fittest" obsolete. Modern medicine means that almost everyone lives long enough to reproduce. Instead of being mighty hunters, we can use Google Maps to navigate to the nearest Mcdonald's to eat 24/7. Our ability to attract a mate is now enhanced with implants, false eyelashes, and photoshopped images on dating sites. The tribes and communities that provided our support systems have been replaced by our Facebook friends.

Now Artificial Intelligence is providing the ultimate tool to complete the transition from biological evolution to technological evolution. This book seeks to tell the story of how technology has caused much of the upheaval we feel, but of how it will ultimately replace the brutality of survival of the fittest with a new paradigm – one where we are not defined by our DNA, but by our dreams.

Table of Contents

Table of Contents

Chapter 1 - The Best of Times

It was the best of times. The year was 1963, John Kennedy was the Commander in Chief, and all was right with the world. We had just kicked the Russians out of Cuba, the economy was great, and Americans were popping out babies faster than Betty Crocker could make cookies.

It wasn't a coincidence that Superman was one of the most popular shows on TV because even middle-class Americans felt just a little bit like Superman. We had beaten the Nazis, split the atom, cracked the sound barrier, and were planning to land a man on the moon. We lived in a country where we knew that anything and everything was possible.

Jack and Jackie Kennedy were the embodiment of the American dream. Beautiful, massively intelligent, powerful, charismatic, and socially elegant. They were the ultimate power couple. To many, Jack and Jackie represented the best of humanity, the ultimate expression of evolution, the Superman and Superwoman of that, or perhaps any era.

And it wasn't just the Kennedys; there were superhumans of all types. Leaders like Martin Luther King, Jr., incredible athletes like Mohammed Ali, Johnny Unitas, and Jim Brown, and astronauts like John Glen were there to show us all that we had the right stuff!

It was a time when Elvis, the "King of Rock and Roll," converted his sex appeal and immense talent into a full-fledged musical revolution. He, along with The Beatles, Bob Dylan, The Beach Boys, Smokey Robinson, Ray Charles, and Joni Mitchell, took music to heights that we've never seen before and may never see again.

It was the same for arts and design. Dali and Picasso expanded our consciousness into new realms, while Andy Warhol combined artistic painting with pop culture. Mid-century homes sported elegant architecture that combined Frank Llyod Wright's design philosophy with a fresh, minimalistic look, and in 1963 Corvette Stingray became one of the most iconic pieces of design America ever produced.

Even literature was at the peak of its creativity. While no-nonsense books from authors like Ernest Hemingway were still popular, Jack Kerouac inspired a new generation with On the Road. At the same time, Ray Bradbury and Isaac Asimov's science fiction novels captured our dream of a future in the stars.

It was, after all, the Space Age, and that was reflected in everything from the taillights of Cadillacs to the futuristic architecture of the buildings being constructed for the New York World's Fair. Of course, the most obvious manifestation of the space age may have been the premiere of the Jetsons in September 1962.

The '60s was the definition of cool. It was the time of Sinatra, the Rat Pack, James Dean, Paul Newman, and Brando. Marilyn Monroe, Bridgette Bardot, and Sophia Loren added an undercurrent of smoldering sexuality to the times. Sean Connery shook (not stirred) this elegant cool and sexuality together to an even higher level in the James Bond movies.

Hugh Hefner converted this latent sexuality into a thriving business with his Playboy empire, which made him a fortune and became a cultural institution. While the titillating pictures gave the magazine its start, people actually did read the articles. Ironically, those articles may have planted the seeds of not only the sexual revolution but also Gloria Steinem's feminist Ms. Magazine. Hefner even combined the Rat Pack cool of the '60s with sexuality to open a series of sophisticated Playboy Clubs whose primary feature was scantily clad bunnies. No matter who you were, these clubs became the "in" place to go. Plenty of paunchy middle-aged men felt just a little like James Bond when they used that secret Playboy key on Saturday night dates with their wives.

As good as the people, the arts, and the culture were back in the '60s, it was science that really stood out. As a kid, I figured that if Superman was busy, the scientists in the white lab coats would swoop in and save us. Back then, we all thought it was a given that we would live to be 120, drive flying cars, and communicate with each other through video screens (well, at least the last one came true).

It was the Space Age, and scientific breakthroughs were coming fast and furious. Our space technology started to bring earthly results with the introduction of the Telstar satellite, which made global communication a reality. In 1961 the invention of the integrated circuit (a.k.a. the chip) made the digital world possible. The "chip" is the core element in computers, smartphones, GPS, streaming video, and all the other enabling digital technologies we have today. More than any other piece of technology, the "chip" changed the world. Of course, once we developed computer technologies, we needed to program them. The '60s marked the debut of the legendary Basic computer language that allowed everyday humans to leverage the power of these incredible machines. It also spawned a new class of professional – the computer geek. Hundreds of thousands of them would eventually program code into almost every device we own. Of course, even your car has millions of lines of computer code these days.

Look at almost any photograph from the 1960s, and you'll see beautiful, well-groomed men and women working industriously and starting families in an upwardly mobile environment.

These photos often embody everything we now think of as the American dream. Houses in the suburbs, white picket fences, kids playing in the yard, and the handsome hero husband coming home at the end of the workday to the impeccably dressed wife with dinner on the table. Ozzie and Harriet, Father Knows Best, and Donna Reed all typified the happy American family. The title of the show Happy Days pretty much said it all – and what more could anyone ask for?

After World War II, the United States and its allies thrived. By the '60s, most of the western world had started to peak intellectually, athletically, militarily, economically, culturally, sexually, and scientifically.

The 1960s may have been the peak of human evolution.

These days we take medical technology for granted. But it's easy to forget that life was fragile before modern medicine. Any small cut could lead to a massive fatal infection. Mold, fungi, and bacteria easily infested people's lungs, causing a slow death at an early age. A simple abscessed tooth or gum disease could cause a person to starve to death. Tetanus was rampant, causing people's jaws to lock up, rendering them unable to eat. Simple broken bones would often not heal properly, limiting a person's mobility and ability to hunt, farm, or work. People went blind from swimming in polluted rivers. Impure water and cholera killed thousands yearly, and then there were plagues and smallpox. Even childhood diseases like measles and mumps took lives because people had no means of reducing fever.

Many babies and mothers didn't make it through childbirth, and about a third of children didn't make it through puberty.[1] Before modern medicine, you had to be either very lucky or very tough to survive. Those born with genetic advantages, such as immunity to the black plague or malaria, typically lived long enough to pass their genes down to their offspring. This was evolution at work.

The '60s also saw an explosion in medical technology. Watson and Crick discovered DNA in the late '50s, which laid the groundwork for so much of what we call medicine today. Blood transfusions became commonplace, enabling long, complex surgeries with high survival rates. It also brought the beginnings of many lifesaving technologies, such as pacemakers. While Penicillin was developed in the late '40s, the '60s is considered the golden age of antibiotic development. The Sabin sugar cube vaccine administered to all school children in 1961 assured that no one in the United States would be crippled by Polio ever again. Since then, there have been hundreds of medical breakthroughs, from bone marrow transplants to gene therapies to COVID vaccines. Science has even been able to mitigate many debilitating mental illnesses like bipolar disorder. For the first time in history, these medical breakthroughs allowed even those with severe medical conditions to survive to reproductive age, mate, and produce offspring.

The FDA approval of the birth control pill on May 9th, 1960[2], made contraception convenient, unobtrusive and reliable for the first time in history. This breakthrough helped make reproduction a matter of conscious choice rather than random chance or divine intervention. The release of "the pill" dramatically changed reproductive dynamics. Together, these reproductive and medical interventions marked a huge inflection point in evolution's ability to continue forward.

Was May 9th, 1960, the beginning of the end of evolution?

In later years, new reproductive technologies allowed people to breed whenever they wanted. Technologies like egg harvesting, in vitro fertilization, and surrogate motherhood allowed anyone, regardless of age, sex, or medical condition to procreate and pass their DNA to subsequent generations.

Biological evolution is all about survival and reproduction. It works simply because the animals with the best survival characteristics live long enough to reproduce and pass on that enhanced survivability to their offspring. Over time the most survivable factors become prevalent while those less survivable traits (and individuals) become extinct.

With a car and the grocery store down the street, you no longer have to be a great hunter or skilled farmer to ensure the family can eat. In modern homes, you don't need to be proficient at making a fire to avoid freezing at night. As our technology and social safety nets continued to develop, we no longer need to be strong, smart, or even sexy to survive. In this new world of technological miracles, almost anyone could make it to reproductive age and reproduce.

But Toto, I don't think we're in the '60s anymore. If you were a follower of science fiction back then, you might have expected humans to have developed extra-large brains by now. This was dramatically portrayed in the 1963 Outer Limits episode The Sixth Finger, which features David McCallum evolving an extra-large brain that grotesquely expanded his cranium. Obviously, that evolution didn't happen. In fact, some would posit that we are devolving. Take a look at any group photo from the '60s, then look at any group photo taken in the last few years. Do the people in the newer picture look more evolved? Instead of enlarged brains, we seem to have evolved enlarged bellies.

> *If the '60s were the young Elvis, we are now*
> *the old Elvis. Bloated, drugged, and sliding*
> *downhill at an ever-increasing rate.*

According to the Centers for Disease Control, middle-aged men are 27 pounds heavier and middle-aged women are 25 pounds heavier than they were in 1960.[3] Diabetes (which would have been fatal a century ago) has reached epic proportions. The Jane Fonda Workout, Step Class, and Jazzercize are relegated to the nostalgia channel. Many of those Peloton bikes have become expensive clothes racks. Any trip to Walmart or Disney will reveal that much of the population can barely walk and that electric scooters are many people's first choice for locomotion.

The two-parent model evolved over the eons because it was the most successful way to produce survivable children. But the Ozzie and Harriet family unit from the '60s is practically gone. Lower marriage rates, higher divorce rates, and mass incarcerations have made single parenthood a new norm – especially in minority communities.[4] It has become so mainstream that many people are now actively choosing single parenthood over the traditional family unit. While being a single parent is extremely popular, there are repercussions for both the children and society.

With parents often absent, role models become even more critical. But these days, there are few good role models that children (or adults) can use to guide their lives. Those public figures we used to hold in high esteem are tragically flawed with alcohol, drug, sexual, legal, psychological, or other serious problems. Most political leaders are seen as self-serving or dismissed as bellicose bullies. Hollywood stars are more likely to be in rehab than on the film set. Even sports heroes like Simone Biles and Mikaela Shiffrin are suffering from severe cases of performance anxiety.

As a society, we have gone from Mission
Impossible to This is Us!

While the parents of the 60s didn't give a thought to having their children play outside all day long, parents these days are afraid to let their kids out of the house. Adding to the already horrendous crime rate in the streets, there is a mass shooting almost every day.[5] Gun violence after car accidents is now the most significant cause of death for young people.[6] Then, there is Fentanyl, the equal opportunity taker of young lives. It's no wonder that most inner-city kids don't believe they will make it to 30.

Evolution depends on successful mating, but low marriage rates and high divorce rates mean the majority of fertile people in the U.S. are not married. A 2020 Gallup poll says that the number of people identifying as LGBTQ+ has almost doubled since 2012[7]. It's no wonder that the birth rate in the U.S. has been cut in half since the 60s is now far below the replacement rate.[8]

Even our Superman, science, is no longer respected. Facts of any kind are considered meaningless by much of the population, and conspiracy theories abound. Noted scientists like Dr. Anthony Fauci are ridiculed and regarded as just ordinary folks whose opinion isn't as valid as a quick Google search. Even evolution is disputed by a high percentage of Americans.

During the COVID epidemic, much of the population rejected lifesaving vaccines, which saved the world from diseases like Polio and Smallpox just a few decades earlier. The anti-vax movement became so prevalent that roughly 70 million Americans chose not to be vaccinated.[9] Many citizens in trusted professions like teachers, police officers, and firefighters decided they'd rather lose their jobs than take a vaccine with impeccable safety statistics. Televised stories showing body bags in refrigerated tractor trailers and deathbed pleas from former anti-vaxxers didn't seem to make much difference.

Does this sound like a species that is trying to survive?

Talk to practically anybody these days, and they'll tell you the same thing – we're going backward. An AP-NORC survey released on June 29, 2022, revealed that 85% of Americans said the country is going in the wrong direction.[10] Physically, socially, intellectually, sexually, and in almost every metric you can mention, we are not the society we used to be. You can blame it on politicians, modern medicine, processed food, technology, TV, or the Internet, and to some extent, all of that is true. But no matter what you blame it on, the current conditions are no longer conducive to humans advancing as a species.

Chapter 2 - The Evolution of Evolution

Whether you believe that God spontaneously created humans in the form of Adam and Eve or that we somehow evolved from primordial slime, the existence of evolution is undeniable. Of course, when we speak of evolution, we think of Darwin and his theory of natural selection. It is a simple theory that in and of itself does not conflict with the idea of divine creation. It states that organisms with the appropriate traits to survive in a given environment will produce more offspring than those who don't survive to reproduce. The offspring of those "fit" organisms inherit those beneficial traits that help their subsequent offspring to survive. Eventually, these organisms become dominant in their environment, while less "fit" creatures may become extinct.

When Darwin visited the Galapagos Islands in 1835, he noticed that birds called finches were present on each island in the chain. As he traveled from island to island, he was surprised to find that each island had its own unique species of finch that had physically adapted to thrive in the conditions of that particular island.

Because of the immense expanse of ocean between the Galapagos and the mainland, Darwin assumed that all the finches originated with a single species of a common ancestor. He theorized that while they had a common ancestor, the finches evolved over time to meet the particular requirements of living on each island.

While Darwinism is often characterized as the "survival of the fittest," it is really about an organism surviving to reproduce any way it can. Fitness might mean that you can win a fight to the death against an enemy. It might also mean being able to run faster than enemies can pursue or learning how to hide whenever a fight is in the offing. Fitness might also mean you can hike 20 miles to find a water source or tolerate vast temperature extremes without dying from hypothermia or dehydration. In some cases, the smaller members of the species might fare better than larger members who can win battles – but require too many calories and fluid to survive in a drought. In other cases, the DNA might survive simply because they are more attractive to potential mates or promotes fertility.

So, no matter how they came to be, there is no debate that all living things exhibit evolution. They must evolve to survive in an environment where the weather, predators, food supply, and competitors constantly change. Want proof? The World Wildlife Fund, conservatively estimates that between 200 and 2000 species become extinct every year.[11] They didn't evolve quickly enough to adapt to their environment and will never be seen again.

We saw evolution at the most basic level during the COVID-19 pandemic. As we remember all too well, the original COVID virus multiplied billions of times. The more it reproduced, the more likely genetic errors called mutations produced new variants with different characteristics.

The original COVID-19 virus spread quickly and made its hosts sick enough to at least keep them at home – if it didn't outright kill them. It eventually mutated into the Delta variant, which spread more easily than the original. When you spread more efficiently, you reproduce more and become dominant in your environment – which is exactly what Delta did. It wasn't stronger, but it was more prolific. Simple math will show that over a few generations, even a slight advantage in reproduction quickly leads to becoming the dominant strain.

The more a species multiplies, the more subject it is to DNA mutations, and sure enough, Delta mutated into Omicron. Omicron quickly became dominant not only because it was easier to catch but also because it made its hosts less sick. Instead of killing its hosts (which usually stops them from spreading it), Omicron was so mild that victims could continue to be mobile and continue to transmit it. Tens of thousands of people with active Omicron unknowingly attended family dinners, weddings, and other events that people with Delta would have been too sick to attend. Keeping your hosts in circulation does wonders for viral reproduction.

Moving up the ladder of life, bacteria are also constantly evolving. Recently, many have mutated to become resistant to their most deadly predator – antibiotics. Over trillions of bacterial reproduction cycles, occasionally, a DNA replication error creates a mutation that makes the bacteria a little less susceptible to a particular antibiotic. Over time, these bacteria survive the antibacterial environment better than the non-mutated bacteria and eventually become dominant. These "super-bacteria" like MRSA have recently become a threat to humans because we have few drugs that can defeat them.

These same principles apply to all life forms. The Smithsonian's Natural History Museum in Washington, DC, had a great display showing how the modern elephant evolved from the woolly mammoth, snakes evolved from sea creatures, and dogs evolved from wolves. It also displayed how man split off from apes to become Neanderthals and then advanced to our current form. Regardless of how life is formed, the museum and countless other sources indicate that evolution is crucial for an organism and its DNA to survive in an ever-changing world. And while individual organisms are expendable, preserving the DNA, generation to generation, is critical.

An organism's primary purpose is to
protect and propagate its DNA.

Passing your DNA on is only possible to reproduce if you are fit enough to survive to reproductive age. Once you hit puberty, fitness could mean your attractiveness to potential mates. Once a mate was selected, fitness could mean enhanced fertility, having a wide enough pelvis to bear healthy children, living through multiple births, or producing enough milk.

Once children are born, fitness might also mean you were brave enough to sacrifice your own life to protect your child from predators. Like Mufasa in The Lion King, we know that most mammals, including humans, are at their most fierce when defending their children – even to the point of sacrificing themselves. Almost all parents would do anything to assure their children's safety and understand at a very deep level that no parent should outlive their child.

The survival of offspring is more important than the survival of the parents.

Salmon, for example, exert so much energy swimming upstream to their spawning grounds that they die immediately after laying their eggs. Some insects take this concept a bit further by actually killing their mates once the deed is done. Once impregnated, the female praying mantis often bites off its mate's head and ingests their bodies to provide nutrients for the developing eggs. Black widow spiders do the same thing to their mates (hence the name).

Of course, one way to counteract natural selection is to have lots of offspring. Many sea creatures have evolved to lay thousands of eggs so that even if most of them wind up as fish food, at least a few survive.

Having offspring means sex - and that starts with the ability to attract mates. Birds sing, rams fight, and other animals do elaborate mating dances to attract a suitable partner. In the animal world, it is often the male that uses distinctive markings and beauty to attract potential mates. Male peacocks present a massive display of iridescent tail feathers to gain the attention of the relatively homely peahens. The DNA of creatures that are not attractive enough, sexual enough, or fertile enough to attract a mate and produce an adequate number of offspring will become extinct.

Humans are a bit unusual because it is the females that are endowed with the beauty to attract mates. Beautiful women like movie star Elizabeth Taylor, who could attract many suitors, clearly had an evolutionary advantage. But, although she had eight husbands, she only had three children. Evolution only works if you become a prolific reproducer.

But just having offspring isn't enough – you have to make sure those offspring reproduce. There isn't much child-rearing for simple life forms like salmon or black widow spiders. Once the eggs hatch, the babies know what to do and, if they don't get eaten, will likely live to reproduce. In these cases, dad is only helpful as a source of protein.

But as animals go up the evolutionary ladder, parents become more and more necessary. A parent's first job is to provide food for their children until they can find and chew food on their own. Their second job is to protect their offspring from predators, and their last job is to teach their children how to survive and reproduce on their own.

In the end, evolution has three rules:
Survive, reproduce, and ensure that your offspring reproduce.

It is important to note that evolution does not have a mind. Evolution is simply the natural consequence of DNA-based life. Evolution does not try to make organisms survive, but those organisms that happen to endure pass those survival traits down to the next generation. Those organisms that do not survive become extinct along with their DNA.

The approach to survival depends on whether you are the predator or the food. If you are the hunter, your strength, skill, and cunning determine if you will survive. If, on the other hand, you are the hunted, you need to come up with strategies to evade or hide from your predators.

The shark is one of the most efficient predators in the ocean. Aside from an incredible sense of smell, sharks have developed an array of sensors that can sense the faint electrical emissions from the sea creatures around them - even in total darkness.[12] Once they spot their prey, they have the speed and power to catch anything that swims. The shark's jaws are not only large and powerful, but their razor-sharp teeth grow in rows which replace themselves as teeth break or become dull. There isn't a better killing machine on the planet.

Unable to match the shark's predatory power, the octopus has become a master of evasion. They hide in small caves at the bottom of the sea and can instantly change their color to match their surroundings. When chased, they emit dense black ink to blind their pursuers. The Netflix show My Octopus Teacher presents an even more fascinating view of an octopus' amazing evasion skills. When being pursued by a shark, the octopus used the suckers on each tentacle to grab pieces of debris from the sea floor to cover its body to shield it from the shark's razor-sharp teeth.

Other sea creatures with less impressive weaponry, like dolphins, have taken to group hunting. They form groups to corral smaller fish into densely concentrated areas so they can swoop in with an open mouth and get a belly full.[13] While simple animals hunt alone, higher animals have found that hunting in groups increases their survivability.

Wolves and humans also use group hunting as a powerful strategy to catch more food. In fact, group hunting may have been the original basis for the evolution of social behavior. Hunting required the ability to coordinate with the other hunters in the pack through body language, facial expression, and eye contact. The concept of leadership naturally evolved from hunting parties because the hunting groups with the best leaders got to eat!

If you aren't a great hunter or there is little game in the area, you have to find vegetables and fruits to eat. Humans, of course, have done this one better by inventing agriculture. Farming differentiates us from animals and maybe even more valuable than hunting in promoting survival. Successful group farming also requires good communication skills. Excluding droughts, agriculture can provide a more dependable food supply than trying to find, hunt, and kill large animals.

It's pretty easy to see how once agriculture became established, the concept of an economy followed. Crops could be traded for meat, farm implements, or seeds for next year's crop. Pelts could be exchanged for arrows or knives. The emergence of an economy takes advantage of the synergy of people doing different jobs and is an immense aid to survival.

It's also easy to understand how an economy led to technology. Better swords, precision arrows, and armored suits are undoubtedly advantageous in battle. The technology of wheeled carts to move goods from place to place or better farm implements helped the community as a whole to survive.

For bacteria, plants, and animals, biological evolution is pretty much their only route to survival. But for humans, the story is different. The evolution of the human brain gave us the intellectual and communication abilities required to create systems to enhance their survival. The early examples were cooperative hunting, agriculture, and the ability to wage coordinated warfare. Later our language skills enabled social and political systems that helped us survive by allowing us to create large cooperative ventures like tribes and nations that enhanced our ability to survive. But it is technology that has truly separated us from the other animals and enhanced our ability to survive.

Much of what has been invented, from spears to refrigeration to antibiotics, has supercharged human's ability to survive on this planet. Our interstate highway system helps us survive food shortages by providing a functional food distribution system. The electric grid helps us survive by providing heat and light. Cell phones help us survive by letting us call for help when we blow a tire driving through Death Valley. Our most complex system, the internet, helped us survive the dangers of COVID by allowing us to order necessities from Amazon.

But, unlike biological evolution, where information is passed to our offspring through DNA, technological information is passed forward through written language, mathematics, and data exchange.

For humans, the evolution of social, economic, political, and technological systems are at least as crucial to our survival as physical strength.

Chapter 3 - The Alpha Brain

Although we're not the fastest runners and don't have the sharpest teeth, humans are pretty much at the top of the evolutionary heap. Despite our physical frailty, we are this planet's apex life form because our brains, rather than our bodies, have enabled us to dominate other species. Our brains can leverage language, strategy, and technology to help us create food, conquer enemies and survive dangerous environments. Our intelligence has made us the fittest of the fit in the animal kingdom and created what we call civilization to support our evolution even further. But our brains didn't develop as continuously as our bodies. In fact, over time, we've evolved three different brains that work together to make us the most intelligent species in the solar system.[14]

The story of our brain starts in the dinosaur era when reptiles roamed the earth. Reptiles are simple creatures - so simple that you can almost think of them as a nose propelled by legs. Their brains are programmed to sniff out food or a mate, head in that direction, and hopefully get what they want.

As some reptile lines evolved into mammals, they developed a second, dramatically more complex brain. This mammal brain was capable of communication through emotional expression, simple sounds, eye contact, and body language. This non-verbal communication allowed social interaction and proved a great advantage to species like wolves that hunted in groups.

When mammals evolved into humans, they developed a third brain wrapped around the other two.[15] This highly developed brain was capable of verbal and written language, and understanding the sequencing of events in time. With the ability to represent ideas symbolically, the third brain could process things like art, math, strategy, planning, engineering, and everything else that makes humans unique.

When the third brain developed, it retained, almost intact, the previous two brains of its evolutionary predecessors. So, our human brains have three distinct areas that correspond to our reptile brain, mammal brain, and human brain, all within the confines of our skull. While these sub-brains serve different purposes, they function together simultaneously to provide an integrated approach to survival.

If you can't make up your mind, it is
because you have three brains!

Our brains evolved from the back to the front. The oldest part, the hindbrain, formed during dinosaur times and connects the base of the brain to the spinal cord. It is concerned with essential functions like breathing, heart rate, motor coordination, eating, and sensory reflexes.[16] This reptile brain's task is essentially to keep the machine running.

The hindbrain is the part of the brain that reptiles (and subsequently humans) use to coordinate the movement of legs to allow locomotion, to balance the body over the legs so the animal doesn't fall over, and to coordinate the forelimbs to accomplish tasks like eating.

The reptile brain is connected to the nose's olfactory bulbs, making smell the reptiles' primary sense. Sure, reptiles use their eyes and hearing, but smell provides the biggest evolutionary advantage. While hearing and sight can provide important information at close range, the sense of smell is like long-range radar. Lower animals typically have incredible senses of smell, and in the case of reptiles, that sense of smell directed them, like automatons, toward food. The algorithm is simple - if the scent is getting stronger, keep moving in the same direction. If the smell is weakening, adjust your direction until it gets stronger. Of course, the same algorithm applies to finding a mate.

The reptile brain is pretty self-contained. As it was the first to develop, it formed few communication pathways beyond the spinal cord and sense organs because there was nothing else to communicate with. Because of this, we generally do not sense that it is functioning. The hindbrain continuously controls your heartbeat, respiration, and motor coordination without you even being aware of it. If you have ever driven home from somewhere "on autopilot" without remembering it, that is your reptile brain at work.

The mammal brain (or midbrain) evolved later and helped modern animals survive after the dinosaurs became extinct. The function of the midbrain is primarily controlled by the Limbic System[17] and quite different from the dinosaur hindbrain. Rather than processing information with neural networks hardwired by heredity, it processes information chemically by releasing various hormones and neurotransmitters. Sitting right in the center of the midbrain is the pituitary, the master gland that along with the hypothalamus instantly signals the release of hormones for everything from battlefield aggression to parental love.[18]

The mammal midbrain processes hundreds of variables in real time and can almost instantly release hormones capable of evoking strong emotions. When you argue with your spouse, that sick feeling in the pit of your stomach is the result of the immediate release of stress hormones. When you have social anxiety, it is because your midbrain is stimulating the release of hormones that cause your heart to race. When you see a new baby and involuntarily break into a smile, it is because your mammal brain has just released oxytocin - the love hormone.

The development of the midbrain provided a tremendous evolutionary advantage over the reptile brain. It allowed mammals to instantly know how they felt about other beings in their environment. Unlike reptiles, who were generally solitary animals and judged everything pretty much by the way it smelled, the mammal brain integrated many inputs to produce the emotions needed to facilitate social interactions. It allowed mammals to judge threats in their environment by their size, appearance, body language, and the sounds they made.

If a deer saw a wolf nearby, stress hormones would immediately cause the deer to invoke its "flight or fight" response and run away. Similarly, if a strange human wandered into a neighboring encampment, the midbrain would instantly evaluate whether he was a friend or foe based on eye contact, facial expression, body language and whether he was making aggressive or soothing sounds. At that point, he would either be considered an enemy or embraced as a friend.

Indeed, perhaps the most significant advantage of the mammal brain was its ability to allow animals and humans to "read" each other before formal language developed. It is the midbrain that enabled animals to work together through nonverbal expression – and it still functions today. Many studies indicate that when a new person enters the room, they are judged far more on their body language and vocal tonality than the content of their words.

Like the reptile hindbrain, the midbrain functions pretty much autonomously, but unlike the hindbrain, you are aware of the midbrain's output through the hormone-driven emotions you feel.

Because of the chemical orientation of the midbrain, this emotional awareness is far more nebulous than hardwired functions. Worse yet, the secretion of hormones and neurotransmitters is extremely difficult to consciously control. Since the midbrain does not directly decode language, we are finding many of the most effective therapies must combine meditation, visualization, and medication with traditional talk therapy.

The newest part of the brain, the forebrain or neocortex, wraps around the midbrain. While many mammals have significant forebrains, it is the most highly evolved in humans. It is responsible for the "higher functions" like mathematics, planning, and language. Unlike other parts of the brain, which come up with generalized solutions like run, hide, have sex, and eat, the neocortex is capable of coming up with more exact solutions, like how much corn do I need to see the tribe through the winter. It is also capable of executing the verbal and written language to let the other members of the tribe to gather and store those ears of corn.

At the very front of the forebrain is the most highly evolved part, the prefrontal lobe.[19] This area enables things like strategy and planning that require events to be sequenced in time. The prefrontal lobe is responsible for engineering, technology, logistics, and predicting future outcomes. It allows us to plan the future by internally simulating and evaluating future scenarios. It is also the seat of our consciousness.

Unlike the hardwired hindbrain and the hormonal midbrain, which calculate and draw conclusions almost instantly, the prefrontal lobe works relatively slowly. While emotions are limited (love attraction, fear, rage, aggression, etc.), complex problems take time to solve. Questions like how much should I pay for Springsteen tickets require numerical answers. Engineering problems like how thick should the beams on a bridge should be can take even longer because they only have one precise answer, and being wrong can be deadly. Because of this, the prefrontal lobe can take from minutes to months to solve a problem.

So, as humans, we are often caught in a quandary between the instant emotional midbrain and the frontal lobe's slow, well-thought-out solution. You feel this as anxiety when your midbrain-generated "gut reaction" disagrees with the "rational" explanation provided by the forebrain.

Because the midbrain works so quickly, most people make most decisions emotionally, because they "feel right". When asked why they made the decision, they unconsciously use the immense power of the forebrain to retroactively generate a logical rationalization for the decision their midbrain has already made.

In fact, the most significant predictor of human success is a highly developed prefrontal lobe.[20] People with this trait can override the emotional midbrain to sacrifice near-term pleasure for the promise of long-term gain. It is the part of the brain that can look down the road ten years and realize that although it is painful in the near term, going to college will pay off.[21] It is also the part of the brain that great leaders use when they envision building a business or a nation.[22] Ironically, it is the people with the most highly evolved frontal lobes that plan their families and have the fewest children.

The prefrontal lobe is responsible for pretty much everything we call civilization.

Unlike the reptile and mammal brains, which are highly structured at birth, the forebrain is structured by experience. The minute a person is born, Billions of neurons start their lifelong programming process by observing and learning from everyday life. Every experience exercises specific neural synapses which resonate when similar situations are encountered. Billions of these synapses are being programmed every second with every sight, sound, thought, motion, and emotion we have.

When a human baby is born, it is completely helpless. Its dinosaur brain reflexes keep its heart beating, its lungs moving, and it crying for food. Its midbrain also begins to communicate by holding eye contact with its parents, decoding facial expressions, and responding by grunting, gurgling, or crying. These behaviors are automatic. They cannot be consciously initiated, as newborns do yet have a sense of consciousness. But, from the moment the child enters the world, its forebrain starts learning, eventually leading to consciousness. Learning is slow at first, but patterns are easily recognized over time, and children start to learn rather quickly.

Mammals learn in two ways: trial and error and imitation. Trial and error is the iterative process that a baby goes through when it learns to walk. It stands, falls, stands again for a second, and then falls again. Eventually, it retains the nuances of balance. But if we had to go through trial and error to learn everything, we would never be able to evolve complex behaviors that contribute to survival.

So, mammal brains are also structured to learn from those around them. They contain specialized neurons called mirror neurons that assist in this process.[23] When young animals watch adults execute behaviors like walking, hunting, flying, making a nest, mating, etc., the mirror neurons imprint those behaviors in their brains. Mirror neurons program the brain as if the watcher were actually doing the activity themselves. These neurons are so effective that very young children often cannot distinguish between their actions and those initiated by family members. This is one reason that early family bonds are so ingrained.

The reason dogs are called man's best friends is that humans and dogs are emotionally in sync. Because our midbrains are so similar, dogs can connect their emotions with humans through eye contact, barks and whines, body language, and physical contact. Like babies, they communicate very effectively without words. The emotional connection is so strong that no language is necessary.

Dogs naturally live in packs and consider the humans they live with to be part of their pack. Being in a pack was essential to the survival of the dog's ancestors - wolves. Hunting in groups gave them a tremendous advantage in bringing down large prey for which individual wolves were no match.[24] A mammal like a deer was too much food for a single wolf, so sharing the kill was a small price for increased predatory power.

But not all wolves in the pack are equal. The leader, called the alpha, is the wolf that the others naturally follow. The alpha establishes its position by fighting or at least displaying clear dominance through posturing until the other wolves show signs of submission. The less assertive followers are called the betas. While there is only one alpha at a time, young betas can take the alpha position if they challenge and establish dominance.[25] While recent articles have observed that in the wild wolves usually hunt in smaller families, causing some to doubt the alpha hierarchy, the alpha behavior is easily observed when wolves (and other animals) form larger groups or share limited areas.[26]

The alpha is usually the most aggressive hunter. He leads the hunt group and makes sure the pack gets fed. Due to his dominance, he also gets to mate with the most desirable female.

If the alpha is the biggest badass in the pack, it is his DNA that you want to propagate.

Domesticated dogs, who descended from wolves, also look at their human family as their pack. The person who feeds the dog is the dog's alpha because whether they hunt down the food or get it from Chewy, they keep the pack fed. In most cases, the whole human family becomes the alpha because dogs are smart enough to realize where their bread is buttered. In return, they give humans love and affection.

While they don't make good household pets, a troop of silverback gorillas provides an example of pack dynamics that is even more similar to humans. This should be no surprise as they are one of our closest evolutionary relatives. Typically, the silverback alpha male sits in the middle of the group, eating and yawning. He is unconcerned with his mate's activities and the antics of his kids. If we sat him in a Lazy Boy and gave him a newspaper, he would look like a hairy version of Archie Bunker. The young gorillas are typically wrestling or chasing each other like human brothers. Once in a while, they'll snuggle up to their dad, whose face spontaneously displays the look of a proud papa. After a minute or two, he will lovingly shoo them away and go back to watching over his domain. Like humans in the 1960s, the male silverback's job was to bring home the bacon, while the female's job was to deal with the kids.

As with humans, adolescence sometimes complicates the parent-child relationship. When the silverback's sons get old enough, testosterone takes effect, and a son may challenge his dad the way human adolescents sometimes challenge their father. Most of the time, a few thumps on the chest and a mighty roar put this challenger back in his place. But, if the son is dominant enough, he may take the alpha role from his father. If the son is unsuccessful and ready to mate, he will often leave the pack to find a mate and start his own household – just like humans.

Gorillas are ferocious animals, especially to anyone seeking to harm them. But within their troop, they are caring parents and have a family life that many humans would envy. When the alpha male gets older, the family lovingly takes care of him and often lets him still think he's in charge. Just like in a human family!

The ancient human alpha was very much the same. Jason Momoa's portrayal of the leader of the barbaric Dothraki tribe in Game of Thrones is a great example. Powerful, fierce, and fearless, he ruled via the double-edged sword of magnetic charisma and fear. Alphas are outgoing people and usually the coolest guys in the room. Think Brando, Brad Pitt, The Rock, and Bruce Willis – they are in control and the type of people who can inspire armies to follow them into battle against all odds. As you would expect, they absolutely hate to lose (since losing often means death). While they are often slow to anger, they elicit fear in friends and foes alike because if you get on their wrong side, they show no mercy. Like Gorillas, these fierce alphas can be loving husbands and wildly protective of their families. And why not? Their main job is to propagate their DNA to future generations.

In his 1982 book *Chimpanzee Politics: Power and Sex Among Apes*, Franz de Wall suggested that the dynamics of the chimpanzee colony could be applied to human power hierarchies. In the case of humans, however, prestige, wealth, influence, and domineering behavior were probably more important to alpha status than sheer physical strength. Studies of apes and humans have suggested that many of these alpha characteristics result from relatively high testosterone levels.[27][28]

In addition to alphas, many psychologists have further classified human males into beta, delta, gamma, and omega personality types. Although each type has unique characteristics, only alphas are natural leaders. Most followers are betas, who are generally competent but less dominant than the alpha. While being a beta might seem useless, it is a tremendous individual survival strategy because everyone is gunning for the alpha. The beta stays in the background and is likelier to die of old age than in a power struggle.[29]

Interestingly, betas have the potential to become alphas under the right circumstances. In The Game of Thrones, the chief of the Dothraki was always surrounded by fierce beta warriors who were just itching to take over if he showed even the slightest weakness. Betas may take over the alpha position by challenging an older alpha, or they may be thrust into the role if the alpha dies in a fight. Either way, the next most dominant beta is usually quite capable of taking on the alpha traits. A beta thrown into an alpha role's brain will respond by unconsciously increasing its hormone levels to the point where aggressive alpha behaviors start to emerge.

While betas are the alpha's lieutenants, other types fill different roles. Sigma males, for example, are highly competent, but rather than function as leaders, they function as lone wolves. A great example might be Clint Eastwood in practically any of his movies. He is a force to be reckoned with, but he works alone and doesn't care about getting the girl.

Omega males are the world's nerds, as typified by the crew in The Big Bang Theory. In many ways, they are the emotional opposite of the unshakable alpha whose very presence inspires people. Many Omegas display nervousness and attention to detail that, while very useful, is less than motivational. They don't get the girl, either.

All of the categories are highly productive in society, with one exception, the delta male. Delta males are broken - either by genetics or a traumatic experience. They are angry, resentful, suffer low self-esteem, and are self-sabotaging. Their internal feelings are reflected in their body language, vocal tonality, and lack of eye contact, which to a large extent, prevent them from productive social interactions. They rarely get the girl, and if you don't get the girl, you don't reproduce.

Throughout this book, we discuss evolution almost exclusively in terms of male behavior and male alpha/beta hierarchies.

There are three reasons for this: The first reason is that we are mammals, and through evolution, we have inherited the alpha male domination that is common to most mammal species.

The second reason is that until relatively recently, males dominated almost all of human history. This provides many examples of alpha male domination for us to study.

The third reason is that men have been endowed with the size, speed, and strength to excel in a "survival of the fittest" world. While the unfair treatment of women is a stain on humanity as a whole, it is, to a large extent, another brutal side effect of how mammals have evolved.

It should be noted, however, that females can be just as effective leaders as males. Literature is starting to emerge, suggesting that females have a somewhat parallel power structure. We've all met those alpha "queen bees" who dominate the room and, in many situations (like most marriages), dominate the male.

Whether you are an alpha, beta sigma, or omega is determined primarily by your hormonal balance. An animal's natural hormonal-driven tendencies determine its place in the social hierarchy. A naturally aggressive individual would likely be an alpha, but an inherently passive individual might be a beta or omega.

All these groups evolved because they were necessary to make an animal or a human pack work. Everyone can't be a leader.

From a survival perspective, multiple leaders create dissension and slow decision-making – which can be fatal in the wild. But no matter how strong the leader is, the followers provide the horsepower to make it happen. And every group could use at least one omega type to take an analytical look at the situation.

In the original Star Trek series, Captain Kirk was obviously the alpha male. But rather than having the typical beta first mate, Gene Roddenberry dared to go where no one had gone before by making an omega rather than a beta second in command – Mr. Spock. Spock was the perfect non-hormonal foil to the overly emotional (and over-acted) Captain Kirk. But together, they made a good team, just as the natural mix of alphas, betas, sigmas, and omegas work today.

The family unit and social hierarchy evolved as it did because it was the best way to survive. But these days, women don't need an alpha male in their life to bring home the bacon. Technology like cars, washing machines, cell phones, and Amazon make child-rearing possible for a single person. With a quarter of American children born to unmarried mothers and over half of the marriages ending in divorce, single-parent families are becoming the rule rather than the exception.[30] [31]

These days the extended family of the watchful neighbors and the wise uncle has been replaced by people who don't want to get involved. Teachers are also more reluctant to take on the ever-increasing problems of teens. It's easy to see that with fewer parental role models, a teenager's peer group can become a much more important (and not necessarily good) influencer. This may be why social media has such a strong pull on teenagers. They're looking for direction.

The question becomes whether this new reality benefits or harms our evolution - and the answer, to some extent, revolves around role models. As we've discussed, children use mirror neurons to internalize their parent's behavior. With only one parent to act as a role model and few good role models in society, we have to wonder if there is enough good behavior to emulate. The problem seems especially acute with male children who lack a resident male role model.

85% of incarcerated youth were raised in fatherless homes.[32]

Chapter 4 - Sex

Sex. If you are a teenager, it's practically all you think about. If you are evolution, it's practically all you are about. Of course, sex creates offspring, but sex also turbocharges evolution. There is a reason the immensely complex mechanism of sex evolved early in the history of life. It is because mixing different versions of DNA through sex enables a species to adapt to its environment far more quickly than it would if it were just creating clones of itself over and over again.

When you learn about evolution, you're taught that random mutations drive the changes that make a species evolve. That's true, but random mutations are exceptionally rare - especially beneficial ones. A case in point is the COVID-19 virus which has infected almost a billion people worldwide.[33] Each infected person probably produced at least a billion copies of the virus in their body.[34] Even with a billion, billion copies created, scientists so far have discovered only five significant COVID variations (although there are many subvariants).

A virus is the simplest life form on the planet, so one stray mutation can make a big difference. If the probability of a mutation is that low in a virus, imagine how tiny the likelihood of a beneficial mutation is in the solar system's most complex organism - a human. The chances of developing a mutation that increases running speed or bestows eagle-like eyesight or the ability to survive for days in the desert without water is infinitesimally small. If it weren't for sex, we would evolve so slowly that we couldn't adapt quickly enough to survive in a changing environment.

The alpha male isn't just about having the best warrior to ensure the babies are fed and defended; it's about having the best (most survivable) DNA in the pack. The alpha male gets the first pick of the females - and the female he picks is usually the one with the best DNA. The combined DNA of male and female alphas gives the species its best chance of surviving and evolving forward.

While the alpha male represents the most dominant member of the pack, the alpha female generally represents the female most likely to bring survivable offspring into the world. With humans, this trait is often signaled by physical beauty. Although different cultures have different beauty standards, they are usually based on cues that indicate the woman is healthy and would be a fruitful mother. Beautiful estrogen-nourished skin, sturdy legs, large breasts for feeding babies, and enough pelvic girth to ensure successful births. Sounds a lot like Marilyn Monroe.

Ironically, in the animal world, it is often the male that is the most beautiful - sporting prominent features to attract females. Only the male lion has the large, luxurious mane. Male birds are famous for their bright colors, and only the male Peacock produces that incredible display of iridescent plumage.

Many species also go through elaborate mating rituals to attract mates. Sometimes it's beautiful birdcalls; other times, it may be intricate mating dances or locking horns with another suitor to win the female's favor. All of this evolved to show potential mates they have the right stuff (a.k.a. DNA) to continue the lineage.

While we have portrayed the alpha as the most formidable male in the pack, it is often just a show - like the Peacock. You see, it's hard to know who the best hunter and the best provider really is, but the male who postures and puts on the best show usually gets the nod. It may be the gorilla that beats his chest the hardest or growls the loudest rather than the one that's actually the toughest. The only way to know is if a challenger fights the alpha for dominance; and in practice, that fight rarely happens. Most disputes are resolved through growling and baring teeth until one of the competitors backs off. When a brawl does happen, it's seldom to the death because the last thing the species wants to do is kill off its best DNA.

Whether they are the best or not, there is research that says alpha males generally have more testosterone than beta males.[35] Testosterone enhances male characteristics like muscle development and encourages many aggressive behaviors that male alphas display. Potential competitors who see these aggressive behaviors assume that this is a guy brimming with testosterone and, therefore, someone you don't want to mess around with.

Although humans have evolved well past gorillas, they still choose their mates in a very similar way. If you're a female, you're looking for a strong masculine male who can ensure that your children survive to reproductive age. We all know there is no more powerful instinct than a mother's drive to keep her children safe.

In the animal world, the alpha male is the fittest
– with humans, it's a bit more complex.

The animal that can best protect his kids in the animal kingdom is usually a skilled hunter and fighter. But in the human world, we can't protect our young by ourselves – we need to leverage the community. We need to live in low-crime areas, send our kids to safe schools, and make sure that the police are doing their job. In this model, the ultimate human alpha might be the guy with the most prestige, the guy who is friendly with the mob, or the guy with the most money - like John Kennedy.

The quest to find a mate starts right after puberty. Raging hormones not only change teenagers' bodies but also change their brains. Not surprisingly, the part of the brain that changes the most is the hormonal mid-brain.[36] Anyone with a teenager knows rational thinking isn't their strong suit.

But, before teenagers are ready to produce offspring, they must figure out who they are and where they fall in the social hierarchy. One of the reasons that social status is so critical to teenagers is that they need to know where they stand in the social order before selecting an appropriate mate. That's why, right after puberty, middle school is three years of jockeying for social position and typically the most socially brutal time in most people's lives.

When their hormones kick into high gear, middle school boys try to impress each other with their athletic prowess, roughhousing, and daring but stupid feats designed to show who is dominant. Conversely, girls determine dominance through thinly disguised vicious verbal attacks on each other. These range from attacking the girl's appearance to their fashion sense, their friends, and their social acumen. An excellent illustration of this is those "mean girl" high school movies, where the arrogant but beautiful, dominant female dates the quarterback and brutalizes all the other girls around her. With the advent of the digital age things have gotten worse, as young people have been given a nuclear arsenal of social media weapons to use against each other. In real life, many victims of social bullying never fully recover.

At some point, the kids break up into groups they can feel comfortable in. The jocks, the geeks, the student council members, the artsy kids, etc. Within each of these groups, however, there's an alpha, a bunch of betas, and some omegas. Any kid hanging out in the wrong group is typically badgered, ostracized, and sometimes even physically accosted until they leave. This teenage drama is not without purpose, as it likely exists as an evolutionary mechanism for determining social hierarchy and status.

There's an old saying that if someone is too weak to fit in a group, the group either wants to comfort them, get away from them, or kill them. As cruel as this sounds, from an evolutionary perspective, it might make sense. Eliminating the weak DNA from the gene pool benefits the tribe in the long run. This tragically explains why teenagers often relentlessly badger more vulnerable members to the point where they end their own lives.

Our mid-brains haven't evolved much since ancient times, and the teenage hormones that evoked hunting/war parties still flow today. The street gangs that often start in middle school are direct descendants of those ancient hunting/war parties that the ancient tribes used to survive. This is why many kids are unconsciously so attracted to gang life – especially in neighborhoods where their survival isn't guaranteed.

Eventually, in high school, the membership of the groups becomes more stable, and the brutality diminishes. Now, as in the ancient hunting/war parties, youth discover the power of teamwork. This is one reason clubs and organized sports are so prevalent in high school and college.

Traveling sports teams for teens are now in vogue. Part of the reason is that being a successful athlete is the equivalent of being an alpha - and almost every parent wants to have an alpha child. While the parents tell you that they are paying thousands of dollars for traveling sports teams so that they will be able to get a college scholarship, they know the odds are against them. It would be cheaper and easier to put time into studying to win an academic scholarship. But, their true motivation may be to prove to themselves that their child is an alpha.

Go to any PTA meeting and listen to the parents talking about their children. If a parent brags, "my son won the state chess championship," the parent of the alpha smirks and proudly says, "my son is the starting quarterback on the football team." There's no contest because the human midbrain hormonally identifies the highly athletic child as an alpha but doesn't produce much hormonal response to intellect.

There are, of course, opportunities to dominate in different groups. There is always the cool kid who dominates because he has one of the trademark traits of alphas - remaining cool. And then there is the 5'4" wise guy who has learned to use intimidation and posturing to dominate his group.

You might ask why is finding a place in the hierarchy so important. Well, in the human world, there's not just one alpha to follow; there are lots of them. The human race is a patchwork of subgroups. It's essential to fit in with a group so that they can provide the support an individual needs to survive. Outliers are pretty much left to fend for themselves.

But there is another reason to find your place in the hierarchy. It helps to determine whom you will mate with. Evolution demands that organisms try to mate with the highest-status individual possible so that the most superior, survivable beings will be produced. In any given group, the alpha male picks the most desirable female; the highest-ranking beta chooses the next most desirable female, etc. The females naturally know where they fall in the desirability hierarchy and set their sights on the appropriate male.

The old cliché about the quarterback hooking up with the head cheerleader isn't just a cliché.

Another way to look at this is that evolution wants to maintain the high survivability DNA that evolved over thousands of years. So, males and females mate with someone at approximately the same physical and social level as they are. If an alpha reproduces with a substandard mate, much of that high-value DNA will be diluted.

Most successful couples are at about the same level, and we question it when they're not. Think of George Clooney and Amal or Brad and Angelina as examples of those alpha matches. A couple of tiers down might be Sara Jessica Parker and Matthew Broderick. A few more tiers down might have been Rhea

Pearlman and Danny DeVito. Look around at married couples, and you see the level matching at work almost all the time.

It's no surprise that Tom Brady married Gisele, a supermodel. But it would be a big surprise if Tom Brady married Roseanne Barr. People would shake their heads. Similarly, when you see Pete Davidson with women like Kate Beckinsale, you scratch your head and say why. It doesn't matter that Pete's clever; our radar is preprogrammed for level matching to be sure that our DNA continues to be optimized.

We've all heard the expression, "she's out of your league," but most of us don't have to be told that a potential mate isn't at the right level. Unlike Pete Davidson, something deep inside tells you this person wasn't the right level for you. It's hard to quantize, but you know. It's that slight uneasiness that occurs when the mix of hormones and neurotransmitters secreted by your midbrain is telling you that something isn't quite right.

That feeling usually keeps average guys from asking out the prom queen. Even if she agrees to go out, the guy will harbor so much deep seeded insecurity that the relationship can't flourish. Practically every episode of the show Seinfeld featured Jerry or George rejecting some charming woman because they were insecure about being at the same level.

Everyone who has been in the dating world knows how hard it is to find a match. And this is another reason that people like to match up with someone from "their tribe." It is far easier to determine someone's status if they lived in the neighborhood or went to the same church. Being from the same tribe automatically helps to set the level - it's like a seal of approval.Love and sex are controlled by the mix of hormones and neurotransmitters generated in the ancient mammal midbrain. Scientists say that what we feel as sexual attraction is a combination of dopamine and norepinephrine stimulated by specific pheromones and visual cues.[37] One day you see, hear or smell someone who stimulates those hormones, and it's all over. You can't think of anything else. Even if you know they are no good for you, your brain is hijacked by hormones, and all you want to do is be with them.

We have proof of how chemically oriented sexual attraction is. In a famous experiment, Swiss zoologist Claus Wedekind demonstrated that one of the most attractive male characteristics was their odor.[38] Women were given photos of men along with their t-shirts to smell. When given a clean t-shirt, the women rated the men moderately attractive. When given t-shirts that had been worn, they rated the same photo as much more appealing. It turns out that the more intense the scent (within reason), the more potent the attraction.

Many have postulated that the women sensed the testosterone that the men had sweated out. Others felt the complex scents were indicative of the man's essential DNA and that the women were unconsciously able to tell which DNA was complementary to theirs. Either way, the unconscious brain makes these decisions, especially in young people whose frontal lobes have not yet fully developed.

Of course, other cues stimulate the midbrain as well. Women may get turned on by a deep masculine voice, physical height, or large, powerful arms. These days, it might even be a sleeve of tattoos that can visually transform even a geek into a badass.

Men might be attracted to the beautiful skin and hair that comes from abundant estrogen. It might be large breasts to ensure the children have an adequate food supply. Of course, breast implants, nose jobs, liposuction, hair color, etc., mean that what you see isn't always what you get. Cosmetic surgery, the right makeup, and hair treatment can help someone who may not have been able to attract someone in their natural state to snag a high-status mate.

Technology has again found a way to circumvent the natural selection process because cosmetic treatments can artificially upgrade a person in level matching for a mate. Unfortunately, these enhancements do not change a person's DNA and don't pass on to their offspring.

Cosmetic surgery is yet another technology
contributing to the end of evolution!

You'll notice that we haven't discussed intelligence as a factor in sexual attractiveness. That is because sexuality is contained in the midbrain, and all the intellect in the world doesn't generally have much influence in the midbrain. In fact, brainy people are often considered un-sexy. Perhaps because talking with them shifts the blood supply from the midbrain to the forebrain. Want to get turned off? Think of math.

On the other hand, alpha bad boys are very sexy. Almost everyone knows a young woman who has been in love with one of those guys that even she knew was bad for her. These men are often arrogant, sometimes abusive, and don't seem to care whether she exists. Maybe it's the smell of testosterone, their posturing, or their looks, but whatever it is, she can't help herself. At the same time, she's totally unimpressed with nice guys who play by the rules and have good jobs. When kids go to college, the frat boys and the jocks get the girls, not the engineering majors.

As people get older, however, things start to change. By age 25, the prefrontal lobe becomes fully developed, allowing people to realize the future ramifications of their current behavior.[39] In other words, people over 25 find it easier to see what would be best for them and their families well into the future. At that point, both men and women can start to override their hormone-saturated midbrain and use their prefrontal lobe to find strategies for producing and raising offspring to adulthood. Suddenly, those unemployed alpha bad boys don't look as good as those high-salary software engineers. Similarly, as men age, they become more attracted to a woman who would make a great mother rather than one who simply arouses them.

In most ancient cultures, parents or matchmakers arranged marriages. This was an evolutionary advantage because elders could see suitable DNA matches without the distraction of hormonal love – which would probably fade anyway. The physical attraction between the bride and groom was irrelevant to the matchmakers because they were only concerned with producing grandchildren with optimized DNA.

In certain ways, we were better off in the days of matchmaking. Judging by the sky-high divorce rate of young marriages, young people may not be great at selecting life partners. A wise, mature matchmaker could easily see when a couple was compatible at the right level without being fooled by posturing, masculine arms, or a curvy figure. Matchmaking helped young people whose frontal lobes weren't yet developed to find suitable life partners. It was undoubtedly a better evolutionary strategy than online dating.

If you've ever been to Aspen for ski season or Monaco for the Grand Prix, you'll notice many successful older men sporting beautiful young women on their arms. These men have every alpha characteristic ; they are wealthy, successful, and influential. Despite their age, the women still see them as more than able to provide for all their needs - far better than most younger men. Of course, men are still attracted to young women for their beauty, youth, and fertility. Despite the age difference, this model still works for producing successful offspring. You rarely see young men with older women, even the rich ones, because men are programmed to look for someone young enough to have and raise children.

> *Marilyn Monroe said it best in Gentlemen Prefer Blondes,*
> *"Don't you know that a man being rich*
> *is like a girl being pretty?"*

Interestingly, gray hair may not just happen when you get older. Gray hair may be nature's way of signaling to the opposite sex that you are now too old to be a successful parent. It turns out that many mammals get gray hair as they age. It makes sense to think that nature would provide some external signal to let potential mates know who is no longer suitable as mating stock.

Attraction and mating is a very animalistic enterprise - controlled by unconscious chemical reactions in our highly hormonal midbrains. But these days, dating is different. People don't meet in person anymore, and online dating provides few of the cues that evolution has used successfully for millennia. Online images give visual clues, but they don't offer eye contact, show how a person moves, express the softness of their skin, or transmit the smell of the hormones oozing from their pores.

In a world of digital media, photoshop, breast implants, fake eyelashes, and airbrush makeup are what potential mates see.

None of it is real, and none of it will be passed on to the offspring. Rather than natural cues to match DNA, online dating relies more on one's marketing acumen and texting ability.

Modern dating does not bode well for
the evolution of our species.

Chapter 5 - The Tribe

Whether you know it or not, your ancestors were members of a tribe. It could have been one of the tribes of Israel, the Inuit of Alaska, the Maasai of Kenya, or any of the thousands of other tribes that evolved worldwide. For ancient people, tribes were the best way to survive in a brutal world.

Tribes provided a tremendous advantage over small families. As opposed to an individual family or a pack, a tribe is sizable enough that it can pursue multiple survival strategies at the same time. For example, a hunting party can be sent out while a second group finds water, and a third group gathers wood. At the same time, people can farm the fields while grandparents watch the babies. A tribe can marshal enough warriors to fight off invading hordes while the women and children flee so the tribe's lineage can continue.

From a genetic perspective, tribes provided the perfect vehicle to enhance evolution. It was small enough to evolve favorable genetic characteristics and large enough to minimize genetic damage from the inbreeding of close relatives.

While early men lived in extended families, inbreeding in a small extended family would inevitably lead to children with genetic defects. This is why most of the world does not allow first cousins to marry. Not only did these genetic defects not move evolution forward, they also significantly reduced the survival rate. In fact, a small extended family would eventually be so riddled with genetic defects that it would almost certainly become extinct. But if the family could grow quickly enough to reach a size of, say, 100 or more people within a few generations, the incidence of inbreeding defects would be greatly reduced.

The tribe was an ideal structure for the continuation of evolution.

Like other mammals, human tribespeople used group hunting to their advantage. But people in tribes also interacted in far more complex ways. While animals can work together on rudimentary tasks, humans' unique ability to communicate through speech and symbols gave tribes the ability to evolve as a social unit.

If the tribe is large enough, it can become even more efficient by creating an economy. Artisans can specialize in crafting tools, kitchen utensils, and weapons in trade for food. Others can sell the food the farmers harvest in exchange for horses or oxen. The ability to divide labor and trade food, goods, and services make the tribe a self-sustaining ecosystem.

Like many mammal groups, human tribes often had a defined territory. A territory with plenty of water and fertile soil was a precious commodity and worth defending because of its survival value.

Highly defensible terrain provides another tremendous survival advantage. If you have ever been to Switzerland, it's pretty easy to see how they have been able to stay neutral for all of these years. The country is entirely surrounded by 10,000-foot-tall mountains called the Alps, which are extremely difficult to cross. Water is another highly defensible feature – even in modern times. The defensive nature of The English Channel, for example, was the only reason the Nazis didn't invade England in World War II. The oceans that surround the United States have also provided America with an unprecedented level of security.

Natural barriers not only keep enemies out - they keep the tribespeople in. This means that the tribe's members can only breed with each other. The result is that each tribe develops distinctive characteristics in the same way Darwin's finches evolved different attributes on each island. As an example, the men of the Masai tribe of Africa have an average height of 6'3", while the African Bambuti tribe average 4'6".[40] Because there is not much diversity in the gene pool, tribes often have a distinctive appearance, body language, and vocal qualities.

But defensible borders are just part of the tribal defense. Animals instinctually recognize and defend against animals from other packs. Neighborhood dogs demonstrate this when they bark to sound an alarm when they see, hear, or smell a stranger. Before the advent of electric fences, dogs would often attack and bite a stranger who didn't heed those warning barks.

Although their sense of smell is not as acute as dogs, humans are exceptionally good at recognizing people from outside their tribe. This critical ability evolved so that outsiders could be identified early enough to mount a defense. For tribespeople, just seeing an outsider instantly releases adrenaline and other stress hormones that initiate the "fight or flight" response. Within seconds, the heart rate elevates, breathing becomes rapid, and the muscles receive increased blood flow to prepare for a fight.

Even today, most of us retain some of these ancient defensive reflexes. Despite knowing better, many people still have involuntary physical responses to people of other tribes. Typically, we don't break into full fight or flight mode - but our stress hormones do elevate, making us unconsciously feel uncomfortable. And it's not just people who look different; we also feel uncomfortable around anyone who is "different" in any significant way.

This effect was studied at Purdue University in 1998.[41] Using student volunteers, they measured heart rates when strangers entered a room. While the heart rate rose slightly when any stranger entered, it escalated a full ten beats per minute when someone of a different race came in. The effect is even further amplified when surrounded by people of another tribe. Try being the only black person in a room full of white people or being the only Jewish person at a church service and see if your heart rate doesn't elevate at least a little bit.

> *This natural evolutionary reflex is the deep-seated root of racism.*

Tribes evolved as an efficient mechanism to protect the tribes' DNA and give it the right conditions to evolve. But along with the physical structure of the tribe, the psychology of the tribe became an essential element in its survival strategy. To do this, most tribes consciously perpetuated the myth that their DNA is superior to that of any other tribe. They typically emphasize in one way or another that they are the chosen people imbibed with unique characteristics by God. Ask the pygmy tribe in Africa, and they will tell you why their tribe is superior to the 6'3" tall Masai tribe. This rationalization helps keep the tribe together and creates a reason not to marry outside the tribe to keep their DNA "pure."

It's no wonder that in the old jungle movies, the African tribes wanted to kill the English explorers as soon as they captured them. They don't want to pollute the DNA they have unconsciously been cultivating for generations. The last thing any tribe wanted was for their daughters to procreate with someone from outside the tribe.

An alpha male led almost all ancient tribes. Even today, people are wired to follow a charismatic alpha with the same blind devotion that a dog follows his owner – regardless of the alpha's competence. While most of an alpha's power comes from their attitude or ferocity, appearance also plays a significant role. In the Disney movie, The Lion King, it is apparent within seconds that Mufasa is an alpha. No lion of lesser status would have that royal bearing, that proud posture, and that mane. In the animal kingdom, powerful males display their grandeur through powerful visual cues like the lion's mane or the peacock's massive plumage.

Whether it came from analogies to a lion's mane, the story of Samson, or just high levels of testosterone, a thick mane of hair has always helped to signify an alpha. Tribal alphas utilized this theme to further leverage their power with elaborate headdresses like those worn by the tribal chiefs of North America, Africa, and South America. Many of these headpieces displayed trophies from their hunts – further reinforcing their alpha warrior status.

The height of these headdresses made the chiefs even more imposing by making them seem taller than their followers. These "crowns" also symbolically kept the ordinary people from standing above the leader.

Of course, the headdress or crown is just one of the ways that alphas display their plumage. Tribal leaders also display their status through opulent residences, trappings, and clothing that display wealth and power. It's no coincidence that the leaders have the most luxurious apparel and residences. They are not just the benefits of the position; they also help to reinforce the perception that the leader stands above the rest of the tribe.

It's tough for even the best alpha to manage the day-to-day activities of a large group. So, tribes naturally developed tools to maintain their cohesiveness. One of those tools was mythology. When you're in a tribe, there are a lot of long evenings around the campfire. There is plenty of time to make up stories about where you come from, what makes the sun rise, the rains come, and the crops grow. There is time to inspire the tribe with stories of battle victories and mythical heroes. These stories are entertaining and provide a common mythology on which the tribe can base its sacred values.

Along with sacred values come plenty of public and private rituals. These devices help keep the tribe controllable by the chief and the medicine man, who typically form an unholy alliance to keep their followers under control. The rituals keep the population distracted and unquestioning. If things are going poorly, the chief can blame the gods and enlist the medicine man to evoke the spirits to make things better. Of course, if things are going well, the chief is the first to take credit.

The sacred values reinforce the behaviors that the tribe needs to survive. The tribe self-enforces these values by ostracizing those who don't comply. Questioning these values and myths was generally not tolerated. We've all seen those movies where a tribe member who doesn't follow the rules is banished from the village to live alone in the jungle. They generally don't survive for very long.

The ability of the tribe to exile members is an excellent evolutionary strategy. Any member of the tribe who was acting poorly or had otherwise defective DNA was typically cast out so their DNA wouldn't pollute the tribe's gene pool. Like selectively breeding a dog species, tribes sought to promote those members who showed the characteristics they wanted.

Sacrifice, especially child sacrifice, was another tool the tribes used unconsciously that promoted evolution. We know that infanticide is practiced in the animal world. But humans also widely practiced infanticide to remove weak or disabled offspring from the gene pool. Unwanted infants were commonly abandoned to die of exposure, and in some societies, they were deliberately sacrificed. While this is typically associated with tribes like the Incas, the Celts, the Mongols, the Egyptians, and many others also regularly sacrificed children.[42]

In addition, almost all tribes had a coming-of-age ceremony that marked a child's passage into adulthood.[43] Right after puberty, boys were typically sent on a mission to accomplish an extraordinarily challenging task like killing a lion. If the boy succeeded, you could be pretty sure he had the kind of DNA that you wanted in your tribe. If he failed, it was just as well that his "inferior" DNA was purged from the tribal gene pool.

Many brutal tribal rituals had significant evolutionary value.

There were, of course, rituals for females as well. Many occurred with menstruation to signal that the girls were ready to bear children. Some of these rituals, like the female genital mutilation are still practiced by some African tribes. One can only believe this procedure was meant to reduce the female sexual pleasure to assure fidelity to her husband. The persistence of this horrific procedure to this day shows how firmly entrenched rituals and sacred values can become, even when they cause tremendous suffering to individuals.

Tribal life almost universally led to belief in God(s). From the Native Americans to the Australian aborigines, nearly every tribe had its God(s). It can be debated whether this belief stemmed from the actual existence of a supreme being or as a natural artifact of the way tribes develop. Either way, religion was a useful construct.

Most people are followers by nature. Without direction, purpose, or guidance, we are just not comfortable being out in the world alone. God(s) were initially invoked to explain natural phenomena like the weather and fertility. But people's inherent need to follow someone or something created a devoted following. God(s) became infallible "super alphas" that people could follow without question because, after all, they were God(s).

Studies show that the desire to follow God(s) and the urge to follow an alpha may be based on the same deep-seated neurochemistry. Research by Dr. Andrew Newberg of Thomas Jefferson University shows that "surrendering oneself to God" suppresses activity in the prefrontal cortex ceding conscious control to the animal midbrain.[44] Other studies indicate that following a religion can reduce cortisol and raise oxytocin levels – just like following an alpha. A study at McGill University revealed that the propensity to follow a religion was inversely related to the testosterone level in men.[45] This coincides with the fact that testosterone-laden alpha males are typically leaders, not followers.

Once there were God(s), there needed to be someone to explain what the God(s) wanted. And that person was the medicine man or shaman of the tribe. Next to the chief, the shaman wielded the most influence in the tribe. But unlike the chief who commanded directly, this influence was always in the name of the God(s). In most cases, the chief and the medicine man worked together in an alliance to use God(s) to leverage their control of the tribe. If the tribe attacked another tribe, the medicine man would be called in to invoke blessings for their victory. If there was a drought, the medicine man would specify what the tribe had to do to persuade the God(s) to bring rain to the region. If a child was dying, the medicine man would mix sacred herbs to try to summon divine intervention to save the child.

From an evolutionary perspective, religion is a valuable tool for keeping the behavior of the tribe under control. It provides standards for how everything is done and, in many cases, practical rules for living a good life. The Old Testament, for example, provided rules for everything from what you could safely eat through the kosher regulations to commercial weight standards to crop rotation. These rules provided a significant survival advantage for the Jewish people.

The other power player who would rise to the top of the tribal hierarchy was the military general. While the alpha chief was assumed to be the best warrior, the general was typically a robust young beta (often a relative of the chief). The general provided a valuable role in defending the tribe from attack but also helped to control the tribe during peacetime. Together, the chief, the medicine man, and the general provided the working political infrastructure to assure the survival of the tribe.

One of the earliest recorded tribes were the Israelites. Led by the charismatic Abraham, this tribe was unusual because they did not live in a defensible nor particularly fertile part of the world. To eat, they became nomads, migrating from place to place. They were held together by a belief in God rather than geographical boundaries.

Eventually, according to Exodus, the Jews would wind up in Egypt, where they encountered a more evolved version of a tribe; a nation. While tribes may be nomadic, most nations have highly defined boundaries.

Whether it is Abraham, Jesus, or Mohammed, every major religion started with a charismatic alpha. Religions, like tribes, desire to go forth and multiply, fear outsiders and believe they are the chosen people. Like tribes, they feel their DNA is superior, so intermarriage is prohibited. As they have no territory, they differentiate themselves by wearing unique tribal clothing like the Jewish skullcap to signal who they are and repel outsiders. Of course, if a religion is to spread, it needs an instruction manual like the Bible to ensure the teachings are followed.

Religions are essentially tribes without borders.

The tribes of Israel dispersed to almost every country in the world, but they retained their tribal characteristics where ever they lived. When the United Nations finally established Israel as the Jewish homeland in 1948, it became a compelling draw for the Jewish people. Like salmon swimming upstream to their spawning grounds, Jews from all over the world feel the need to make an Aliyah, a trip, to the legendary Jewish homeland, despite having established satisfying lives elsewhere.

Perhaps it is the divine spirit, but religions seem to endure far beyond the reign of any individual leader. Rather than the mercurial policies of a mere tribal chief, the teachings of an immortal God give followers the stability they need to pass those sacred values from one generation to the next. But religions, like tribes, would not have flourished unless they helped their followers to survive.

Chapter 6 – War

The haze was heavy in the jungle. It was nearly sunrise, and the sky was just light enough for the trees to form menacing silhouettes on the horizon. Although hundreds of warriors had been gathering in the bushes for the last hour, the only sounds were the first few bird calls of the new dawn. With an almost imperceptible signal from the leader, the warriors slowly began to move into the enemy's encampment.

Silently entering the settlement, they stealthily overpowered their sleeping foes by snapping their necks before they could sound the alarm. Moving further into the colony, they began to search for their ultimate prize - the secret chamber where the queen's children were hidden. With the guards now awake, hundreds of defenders streamed into the tunnels to engage the enemy. The fighting was quick but ferocious, leaving the bodies of most of the combatants strewn in contorted positions on the ground.

Ultimately, the invaders prevailed, carrying the queen's children off to certain death. You see, this was more than a territory war. It was a war that would determine which tribe's DNA would dominate the region for the rest of time. This has always been the way with ants.

We often hear that war is a purely human endeavor and that only humans kill their own kind - but that is not true. Even ants make war, and those wars can be brutal as any human conflict. I remember how fascinated I was watching an ant war in my backyard when I was about seven. Hundreds of ants engaged in mandible-to-mandible combat, ripping arms and legs off, and crushing thoraxes as if it was an episode of Game of Thrones.

And it's not just ants that make war. Many species make war on competing tribes, including elephants, rats, snakes and chimpanzees. Chimpanzees, are genetically one of our closest relatives. The Netflix program Chimp Empire shows that how close we actually are. Each chimp featured in the series is clearly an individual who expresses the full gamut of human emotions through their facial expressions, body language and social alliances. The resemblance to human social interaction is uncanny and after watching the series it is hard to argue that chimps are not sentient beings.

Between 1974 and 1978, two tribes of chimpanzees conducted a four year war in Gombe Stream National Park in the Kigoma region of Tanzania.[46] Other chimp tribes practice a more refined form of politicized warfare that creates fewer casualties while still taking its toll on the enemy. In recent times chimpanzees have been observed gathering in groups to attack an alpha who, while individually stronger, could not prevail against the overwhelming odds. Surprisingly, this power grab targeted members of their own tribe! You could call it a bloody coup or a group assassination a la Caesar. Still, the chimps were smart enough to figure out how to take their targets out without putting themselves in significant danger.[47]

At one time, it was thought that wars started out of desperation. For example, if there was a food shortage, and the choice was to starve, or go to war with a neighboring tribe, that war would be preferable. Similarly, if a tribe was growing too big for its territory, it might need to make war on its neighbors. But the truth is that wars are often just caused by simple aggression. A tragic case in point is Vladimir Putin's war against Ukraine.

This aggressiveness pays off from an evolutionary perspective because the victorious tribe perpetuates its DNA while relegating the losing tribe's DNA to extinction. So, not only does the victor get the spoils, but they ensure that the losers do not have offspring to continue their genetic line.

While war is not common in the animal world, it is pervasive in humanity. There has never been a time in human history when war didn't exist. Even before the white man entered America, the native American tribes had many brutal conflicts with each other – even when resources were adequate. Human wars were often about more than survival; they were about the propagation of the victorious tribe's DNA.

In modern times, we decry war for its unnecessary cruelty, but war evolved for a reason. It is the ultimate expression of the survival of the fittest. In the world of evolution, waiting for predators, bad weather, and food shortages to cull out the weaker members of the herd can take generations. War removes the vulnerable members far more quickly. It also offers the additional benefit of giving the victor additional land and resources. If there were no evolutionary advantage to war, tribes would have evolved to become good neighbors rather than lose citizens through unnecessary killing.

War is one of the most effective ways
to cultivate "superior" DNA.

There are two ways to wipe out a competing tribe's DNA. The first is to kill all the members of the competing tribe. This time-honored tradition of destroying a particular tribe or ethnic group is known as genocide. This horrific practice still exists in modern times, from the Killing Fields of Cambodia to the ethnic cleansing in Ukraine. Of course, the most systematic example of genocide was Adolf Hitler's construction of death camps explicitly designed to eliminate the Jewish race. Hitler was so obsessed with the complete extermination of the Jews that when the tide of the war turned to the allies, he switched to using poison gas to accelerate the process. Hitler wanted to ensure that the Jewish DNA was completely eliminated before the war was lost.

The second method of eliminating a competing tribe's DNA is to be sure that the defeated tribe does not continue to create genetically pure offspring. So, frequently the victorious tribe would not only kill the men but rape the women. This happened with such regularity that it must be viewed as more than a recreational activity for the conquerors. From an evolutionary perspective, inseminating the women of the conquered tribe is a way for the conquerors to further propagate their DNA. Unfortunately, this barbaric practice still happens to this day. During the 2002 Ukraine war, there were numerous reports of Russian soldiers raping Ukrainian women. Some victims quoted the soldiers saying, "We want to wipe out the Ukrainian race."[48]

While humans are animals, many things set us apart. One of the key differences between humans and other species is that humans developed brains with highly evolved frontal lobes. This part of the brain enables planning, engineering, creating strategies, and simulating future outcomes. It is the part of the brain responsible for everything we call civilization. The advent of civilization meant that tribes had many more powerful tools at their disposal.

It's no accident that cities were one of the first hallmarks of civilization because cities dramatically improved a tribe's chances of survival. Those tribes with enough frontal lobe intelligence to engineer functional cities had a far higher chance of survival than those who remained in mere encampments.

Most cities were built in naturally defensible settings. One of the best examples of a defensive city was Masada. Located in the parched deserts of Israel, Masada was constructed on top of a tall mesa, giving it a genuinely defensible position. The location over 1400 feet above the desert enabled its residents to see enemies approaching days in advance. Its steep slopes presented a formidable natural barrier that could slow would-be attackers to the point that they were easy targets for a defending archer's arrows.

Initially designed by king Herod's engineers, Masada had perhaps something more important than a defensible position. It had a cistern system capable of storing enough rainwater to allow the inhabitants to survive for years without rain. In 73 AD, this cistern system, along with a large cache of stored food, let the residents of Masada survive an intense attack by the Roman armies for the better part of a year.[49]

Similarly, many Native American habitats were built atop mesas. This provided an ample flat living space while providing a high vantage point. The sheer natural cliffs were almost impossible for enemies to scale. As time passed, Native American tribes also started to build cities into the mesa's cliffs, providing natural cover from the rock outcropping and giving enemies even more limited accessibility. If you visit Mesa Verde in Colorado, you can see one of these elaborate cities carved into the side of a large cliff. Visitors can climb from one dwelling to the other via wooden ladders constructed of natural sticks and rope the same way the original residents did. These ladders provided relatively easy vertical accessibility but could be removed when the city was attacked, making it nearly impossible for enemies to access the structures.

For tribes that did not have the benefit of natural defenses, they could use technology to build walls. One of the earliest examples was Jerusalem, where high walls were built to protect it from invading hordes. The walls were so crucial that they were made ever higher by almost every civilization that occupied the city over the millennia.

Technology was starting to replace
strength as the key to survival!

The survival value of technology wasn't just limited to making defensible cities. It infused every element of life. Technology allowed artisans to create the tools like hammers, anvils, nails, chisels, and saws that further accelerated the creation of civilization. These tools enabled them to quarry stones, erect structures, and build wheeled carts to distribute food and goods throughout the city. Technology helped to create effective weapons like spears, swords, shields, and later things like crossbows and catapults that clearly aided the residents' survival.

In the eleventh century, walled castles started to become prevalent in Europe. They had the same features as the early cities, but with water more available, many built moats to make the walls even more imposing. Because the middle-ages were marked by an almost constant state of war, these castles were essential to the tribe's survival.

Castles are extremely expensive to build, marking a shift in the definition of alpha power. You see, even the most potent alpha warrior couldn't win a battle against a heavily fortified castle with an army of hired warriors to defend it. Now power was not as much a matter of a warrior's physical power but rather a matter of wealth. Since money could buy castles, weapons, soldiers, and influence, money became more potent than sheer physicality.

In the civilized world, money is more important than brute strength!

The TV show Game of Thrones was a reasonably accurate depiction of the brutality of the Middle Ages (except for the dragons). Each competing warlord had land, a castle, warriors, money, and influence. Alliances were formed, wars were fought, and heads literally rolled. Knights sometimes fought to the death for just entertainment. If you stole something, they would chop off your hand. If you spoke against the king, your tongue would be cut out. For more serious offenses, your arms and legs might be tied to four horses which would be sent galloping in opposite directions ripping your body into quarters. Between the brutality, plague, and famine, the Middle Ages were indeed a time when only the strong survived.

Eventually, through wars of consolidation, most of the warlords coalesced into strategic monarchies with an influential king and landed gentry. Money and influence ruled while knights or their equivalents served as the surrogate alpha males for the purposes of warfare.

Once civilization took hold, wars were no longer just about survival. Instead, they were about politics, religion, wealth acquisition, ego, or just plain genocide. The Crusades, one of the most brutal campaigns in the planet's history, killed millions to spread the word of Christianity.

There are many well-known examples of alpha leaders who went on unnecessary conquest missions. The barbaric Vikings went on countless voyages to conquer northern Europe, Iceland, and parts of North America. Leaders like Hannibal and Alexander the Great sought to conquer much of southern Europe, Asia, and the Middle East. Later, people like Napoleon and Hitler also sought world conquest. Interestingly, many of these conquerors were small in stature. Leaders didn't have to be fierce warriors in civilizations where they didn't have to do their own fighting. Like the queen ant, they had large armies for that purpose.

Instead of size and strength, these conquerors ruled through influence, inspiration, intimidation, greed, and aggression. Their superpower was the ability to walk into a situation and become the boss – like Marlon Brando in The Godfather. These leaders also knew that there were always aggressive betas in their midst, gunning for them. Therefore, their first priority had to be to exert enough influence to keep their armies and political constituencies loyal to them. If not, they would wind up like Julius Caesar and the chimpanzee - assassinated by their own people.

The alpha had now evolved beyond survival – to conquest!

In recent CRSPR gene-splicing experiments, scientists genetically altered hamsters' sensitivity to the hormone vasopressin. Changing this one gene sequence made normally docile hamsters into tyrants. They bullied the other hamsters in the habitat - pushing them off the hamster wheel and taking their food. This all occurred in an environment with plenty of food and space.[50]

The drive to conquer may indeed be where the evolution of the alpha may have overshot. Perhaps these individuals had imbalances of vasopressin and/or testosterone that made them into conquering bullies. Millions of enemies (and followers) have died to satisfy the hormonal excesses of tyrants. While you can say that the better army has superior genes and deserves to survive, tyrants believe those "superior" genes belong to just the leaders. This is why royals consider their lineages so important – because that is where the bulk of DNA cultivation takes place.

Just as with ants, warriors and workers are expendable.

Even without a formal education in genetics, people's animal brains have always intrinsically understood that the "best" parents produce the "best" offspring. In the case of royal lineages preserving the superior DNA was of the utmost importance. The show Game of Thrones was a survival of the fittest competition to determine which line of DNA would win the throne. This is why royals are expected only to marry royals – even if they had to import them from other lands.

Is conquest a good evolutionary strategy? Does it help the entire tribe survive, or do constant wars decimate the population to the point where survival is jeopardized? All we know is that many of the ancient conquering civilizations no longer exist.

Chapter 7 - Nations

The American Revolution not only ushered in one of the most profound changes in governing in the history of the world, but it also changed the course of evolution. Until the American Revolution, with a few exceptions, almost everyone lived under the direct auspices of an alpha in the form of a tribal leader, warlord, or king. It was the way people had lived since the time of the caveman.

But the founding fathers changed all of that. Rather than leaving governing to a royal alpha with "more refined" DNA and a purportedly closer connection to God, the Constitution encouraged people to determine their own fate. Now anyone, regardless of lineage, could become a leader, and citizens could decide their future by voting. Theoretically, at least, we were all now equal.

The revolution started because the colonies did not want to pay onerous taxes to King George III. Declaring independence was the best way out of the paying and a pretty good bet, considering the British army had a limited ground presence in the colonies.

But what started as an economic intervention became an inspiration to the world.

The Declaration of Independence did more than make the United States a sovereign nation; it started a movement from monarchy to democracy that continues to this day. Since the American Revolution, monarchy after monarchy has fallen. The few that remain exist more for their colorful pomp and Disneyesque appeal than to govern.

The U.S. Constitution marked the rise of the rule of law over the rule of man. Rather than governing by fiat as kings and tribal chiefs tended to do, the Constitution created a formal basis for the rule of law. It was specifically engineered to limit the power of any individual by establishing three branches of government that acted as a check and balance system.

Of course, the American Revolution wasn't the world's first revolution, and it wasn't the first democracy. The Greeks invented the idea hundreds of years before Jefferson. But the colonies' revolt against the omnipotent king of England was the act that set the world on fire.

You might wonder what the American Revolution has to do with evolution. The answer is that the American Revolution took power away from a long line of genetically inbred autocrats. From the time of the ancient Egyptian pharaohs, the ruling class considered themselves genetically superior to the populations of their countries. In their minds, and the minds of their subjects, these rulers' genetic perfection made them closer to the gods. Consequently, they were only allowed to breed with other royals.

To some extent, it didn't matter where the royal DNA came from as long as it was from a highly cultivated royal bloodline. Many people believe that the king or queen of England are a direct descendant from a lineage dating back thousands of years. But this is not the case. Years ago, England found itself without a continuous line of royal DNA. So, they imported royals from Germany with the family name Saxe-Coberg. Due to anti-German sentiment, in 1917, the monarchy conducted what was essentially a branding exercise to determine a new name almost in the way the Washington Commanders created a new name for the Washington Redskins. The resulting name was – The House of Windsor.[51]

Much of the time, royal matches were not only "genetically engineered" but also arranged to have strategic value to both royal families. A marriage might be made to bring two countries into a permanent military alliance or to assist with trade. The fact that intermarriage between small groups of royals often led to genetic defects was ignored by these families, who were convinced that their DNA was too highly evolved to mix with mere commoners.

*Democracy limited the influence of
the genetic alpha bloodlines.*

But as you remember, an alpha is not the only one who can lead; just the one in power at that moment. In nature, alphas lead while they are strong, but there are always betas just waiting to take their position. Once these betas usurp the leadership role, they develop even more alpha characteristics. This occurs in humans because the alpha/beta hierarchy is, to some extent, situational. The person who is in charge happens to be the most dominant in that particular group of people at that moment. Put in a different group, however, that person may become a follower with little influence or power.

While leaders innately tend to be alphas, many humans are somewhat flexible and can take on different roles as needed. A 2017 American Journal of Pharmaceutical Education article entitled Are Leaders Born or Made concluded that there is a significant genetic component in the 10% of people who are natural alpha leaders.[52] It also posited that another 20% of the beta population, while not born leaders, could take over leadership if needed. We regularly see this on the battlefield, as soldiers regularly step-up to replace fallen leaders. The same is true for businesses and families. Leaders regularly leave business positions through promotion, resignation or dismissal just to have the next-in-line replace them. In the TV show Shameless, the father is portrayed as a useless drunk with many children. His daughter becomes the leader of the family because there was simply no one else to do it.

Hormonal and behavioral changes naturally occur with new leaders that help them master their new roles. If you have been in the workforce, you've probably seen significant behavior changes when a young person gets promoted into a leadership role. Even relatively meek people automatically stand more proudly and display leadership body language. Their vocal tonality changes, often speaking with deeper, more authoritative voices, and they often become more directive. They also frequently change their clothing to increase the look of authority and grandeur. Sometimes you barely recognize them.

Adopting the Constitution meant that the rule of law was the successor to the rule of man - specifically the king. It's easy to understand why this was an attractive concept. Kings and queens were notorious for making mercurial pronouncements. The famous refrain "off with their heads" is not just something from the movies – it was real. Monarchs were the judge, jury, and executioner for anyone accused of a crime. They also created laws, levied taxes, allocated land, and elevated people to positions of power by fiat - often ignoring their counselors. They waged war, sponsored exploration, made treaties, and initiated diplomatic relationships whenever they felt like it. There were no checks or balances. Those who chose not to follow were dealt with harshly, and often their DNA was eliminated.

The Game of Thrones depicted some of the bloodiest, most brutal scenes ever shown on television. The director was not just pandering for audience share – that's the way things were. In the Middle Ages, the best way to eliminate enemies was to kill them – sometimes in incredibly brutal ways. The rule of law meant that even the king couldn't kill his enemies without a trial.

Monarchs and the landed gentry didn't have to fight themselves – they used their fortunes to hire warriors to fight their battles for them. Knights were, of course, the best fighters. They served as surrogate alphas when the kings, bloated from years of pampered living, were no longer physically competitive. Once civilization hit, wealth became the dominant form of alpha power. Rather than having physical abilities themselves, the wealthiest families could buy the best warriors. Often, knights fought to the death to solve disputes by being the "champion" for a fat, lazy king who could never prevail on his own.

Under democracy, the collective DNA of the nation mattered instead of the DNA of the royal family. If that nation could survive and prosper relative to other countries, its DNA would become more prevalent. This may seem abstract until you consider the survival rate of children in the U.S. versus children in third-world countries. Indeed, it is far easier to survive in the most powerful country in the world than in the developing countries of sub-Saharan Africa.

*A nation's success helps ensure the
survival of its resident's DNA.*

Collaborating to create laws was a huge step up over a singular king. Fashioning legislation with representation from each state theoretically produces wiser and fairer policies. A judicial system with impartial judges keeps the other branches of government from running afoul of the Constitution. Having a president elected by the people puts solid checks and balances on easily abused power.

The U.S. constitution formed structures to create and carry out the rule of law, and many people were required to fill those roles. We needed a president, legislators, judges, and bureaucrats to administer the rule of law. Leadership in the new country was determined by election rather than the DNA of their ancestors.

Most of the people who formed our new government were the alphas of their day. No longer necessarily the fiercest warriors, people like Washington and Jefferson were influential, successful, wealthy, well-spoken, and carried themselves with gravitas.

Ironically, voters tend to elect the people they would follow anyway – the alphas.

The founding fathers hoped that the structure of the U.S. government would lead to a highly collaborative government. But like in the Middle Ages, people who aren't powerful enough on their own form coalitions that can bolster each other's power. In the United States, these coalitions are called Democrats and Republicans.

Al Gore was one of the most qualified presidential candidates in history. But the population perceived Gore as a beta intellectual nerd rather than an alpha leader and elected George Bush, Jr. The Republicans did a brilliant job of nullifying Gore's qualifications by taking deliberate measures to create a Ronald Regan-style "cowboy" image for the junior Bush. They started photographing Bush in outdoor clothing even though he typically wore a suit to work in the city. The famous Bush ranch near Crawford, Texas, was deliberately purchased just before the presidential campaign for the specific purpose of creating a more alpha image for George Jr.

Don't overlook the fact that Bush, Jr. was the son of American royalty, George Bush Sr., and the very queen-like Barbara Bush. Judging from the immense popularity of Queen Elizabeth, it seems like many people still love the royal fairy tale. Could it be that we are still unconsciously buying into the genetic transmissibility of royal leadership?

John Kennedy had all the alpha attributes. Good-looking, charming, a brave war hero, the son of a rich and powerful family, an inspiring speaker, and married to a beautiful woman - he checked all the boxes. As a child, Bill Clinton considered Kennedy his hero and did all he could to emulate him. Clinton was also larger than life - good-looking, charismatic, confident, and charming. People naturally wanted to follow him. Almost everyone who met him in person (especially the women) said that he had an almost hypnotic effect that "made you think you were the only person in the room".[53] While there were always allegations of wrongdoing, like most great leaders, nothing stuck to him, at least until Monica Lewinski. Like many alphas, the opposite sex became his downfall.

Bill Clinton's dalliance should have been no surprise to anyone. Having affairs has always been something that was part of being a powerful leader. Most palaces were specifically built with secret passageways to allow the leader to sneak in his consorts. The Palace of Versailles in France proudly points out these secret doors and rooms out when you take the tour.[54] It's even said that John Kennedy used to sneak Marilyn Monroe into the White House through an underground tunnel.

If you think of these leaders from an evolutionary perspective as alpha males, their job is to propagate their DNA – and females are attracted to them for that very reason. It doesn't make it right, but it does explain why men in positions of power tend to have affairs.

They say that opposites attract, so it should be no surprise that Hillary was, in many ways, the opposite of Bill. Because she was not a natural alpha leader, she lacked the people skills, sex appeal, and charm to attract a passionate following. Unlike Bill, who was like Teflon, everything stuck to Hillary, from Benghazi to her private email servers. Even from the start, people sensed that Hillary didn't have the natural alpha qualities needed to be electable.

It wasn't that she was a woman – if anything, the support for a potential first woman in the White House kept her in the ball game. Most people are thrilled to follow a woman if she is perceived as an alpha leader. Even in ancient times, citizens would follow a charismatic queen like Cleopatra with the same zeal that they would follow a man. In modern times, Golda Maier, Margaret Thatcher, Indira Gandhi, and Angela Merkel were all considered true leaders. Hillary was just not perceived that way.

We've seen throughout history, singular leaders with unlimited power typically abuse that power. One of the best dramatizations of how absolute power corrupts is from Game of Thrones. Near the beginning, Daenerys Targaryen becomes queen when her husband, the king of the Dothraki, is killed. She assumes his place as leader of the tribe, but rather than being a heartless warlord, she earns everyone's respect by ruling with compassion.

Initially, she appears to be a great leader and liberator. Her noble mission is to free those that have been enslaved by wealthy families throughout the seven kingdoms. But, over time, she begins to think of herself as God-like and infallible. She starts to ignore and then even punish advisors who disagree with her. Next, she declares herself ruler for life. Eventually, she becomes even more brutal than the warlords she replaced, ultimately destroying entire cities and burning thousands of civilians alive. This, of course, is the standard operating procedure for tyrants like Vladimir Putin.

Chapter 8 - Capitalism

If America has a claim to fame aside from democracy, it is capitalism. Although capitalism has been around for millennia, the wild success of the American form of laissez-faire rock'em - sock'em capitalism captured the imagination of the world. Starting with Henry Ford's manufacturing breakthroughs, America has always been the acknowledged leader in capitalistic endeavors. The reason is the combination of a government that encourages free market capitalism and the entrepreneurial spirit of the American people.

Capitalism itself is an evolutionary concept. It's survival of the fittest in the world of business. Like animals, companies need to evolve to survive in an ever-changing environment. There is competition for capital, inputs, customers, and labor. There are predators in the form of other firms who want to put you out of business. Thousands of companies who are unable to compete in the corporate jungle become extinct yearly.

Like animal cubs, young companies are especially vulnerable. Without knowing where to find customers to feed them, and a healthy layer of balance sheet fat to see them through the lean times, most businesses fail within the first five years.

When the environment changes, even mature companies may not evolve quickly enough to stay alive. Think about Polaroid, whose innovative instant photographs owned the market – until digital photography took over. Or consider Sears, whose hard copy catalogs set the standard for remote purchasing – until Amazon came along.

> *Those businesses that didn't evolve with the*
> *changing environment became extinct.*

On the other hand, alpha businesses that dominate can spread their "corporate DNA." Starbucks has spawned locations on practically every corner of the country. Apple has put smartphones into the hands of the entire world. After destroying competitors like Barnes and Noble, Amazon used its corporate DNA to dominate in practically every product category you can think of. Despite its reputation as a corporate predator, Amazon is the embodiment of alpha characteristics. In business, the fit survive, but the really aggressive ones, like Amazon, conquer the world!

In the same way, the United States is the alpha of the world economy. That dominance means money, and money buys power. Wealth enabled the U.S. to purchase the weapons, transport the troops and create the atomic bomb that won WWII.

Sometimes money alone can be even more powerful than the military. Our wealth contributed to the break-up of the Soviet Union when they couldn't compete with the United States' massive weapons spending under Regan. The much-heralded U.S. brokered peace between Israel and Egypt wasn't diplomatic genius, it was accomplished by paying $4 billion to those countries to make and keep the peace.[55]

Since the early 1900s the United States has been the world's 800-pound gorilla, but like all good alphas, there is a beta nipping at its heels. China is devoting all its energies toward surpassing the U.S. in technology, infrastructure, world trade, and even defense. It's just waiting for the U.S. to show some weakness before it pounces and takes its place as the world's alpha. You see, just as with animals, nations that don't adapt to their environments lose their spot.

But Capitalism was not the only factor in America's ascension to dominance. The other factor was our people. It is a nation of immigrants. The people who showed the initiative and the fearlessness to immigrate to a new land are most likely alpha individuals.

Betas who obey the rules usually stay in their own countries.

I recently talked with a person about how her family came to America. She told me her father was an officer in the Nicaraguan military when the revolution occurred. Knowing he would be targeted for retribution, he stole a military helicopter, picked up his family, and flew as far north as possible. Once he ran out of fuel, he landed, and his family trekked northward until they entered the United States. He made a good life for himself in the states. This man was a leader, not a follower, and like many people who pick up and leave their nation for the promise of a new world, they believe they are fit enough to survive.

It's not just the immigrants. Americans have always had a dream. Maybe it was that house in the burbs, the fancy car, or membership to the country club. But, in its most basic form, the American Dream is about ensuring that your kids and grandkids prosper – just like evolution. Work hard, and your children will fare better than you do. Think of the thousands of parents who held three jobs, cleaned houses, or dug ditches to achieve that dream.

Within the context of the American Dream, a new form of alpha emerged - the American business leader. And while physical size was no longer a factor - many of these leaders were larger than life. Leaders like Andrew Carnegie, Henry Ford, and Howard Hughes were true leaders who inspired immense loyalty from their followers. They were also ruthless when it came to their competitors. Men like this weren't afraid to risk their fortunes on new ideas and crush their enemies in the process. Their alpha presence was felt every time they walked in the room, leaving their followers inspired and their enemies quivering with fear.

*Business leaders had the same biological
makeup as tribal leaders,
they were just living in a different jungle!*

This model was popularized in both literature and the culture. In the late 1950s, Ayn Rand's Atlas Shrugged portrayed the merits of objectivism, a Hunger Games philosophy of screwing everyone else before they can screw you. This philosophy also proved to be a hot topic in Hollywood. Michael Douglas' character in the movie Wall Street coined the phrase "greed is good." Movies and shows like The Wolf of Wall Street, Succession, and Billions idolized smart, savvy businesspeople who can justify any action required to make a buck. In these portrayals, the companies' mission was to leave the corpses of their competitors on the battlefield to be picked over by bankruptcy lawyers. After all, business is war!

Ayn Rand was right about one thing - American business tycoons hate to lose. In the mid-sixties, Henry Ford II spent almost 100 million dollars (nearly a billion in today's dollars) to defeat Ferrari in the iconic 24 Hours of Le Mans. This incredible effort portrayed in the movie Ford versus Ferrari was based primarily on Enzo Ferrari's comment to Henry Jr. that he was "no Henry Ford." The film is an excellent depiction of the dynamic between two alphas - Henry Ford II and legendary race car designer Carrol Shelby, neither of which could tolerate losing. You leave with the feeling that had Ford not won the race in 1966; they would still be spending money on it to this day.

The modern equivalent of Henry Ford, Sr. was Steve Jobs. Jobs personified all the usual alpha attributes. He was creative, fearless, intimidating, mercurial, and didn't care what others thought. Like many of his alpha predecessors, he pushed his people to the limit and could often be a bit of an asshole. While Henry Ford led the world into the freedom of cheap transportation, Jobs showed the world the freedom of digital technology.

Not all the ideas were Jobs. The Macintosh operating system, for example, was derived from an operating system developed at Xerox's Palo Alto research center. But Jobs offered something that only the best alphas can – inspiration. He led a devoted following of brilliant beta engineers like Steve Wozniak, who made the magic happen. But it is Jobs, not Wozniak who is revered to this day.

The success of business legends continues to inspire everyday Americans. Sporting a business suit instead of armor and an MBA instead of a sword, businesspeople use smarts, cunning, and spreadsheets to succeed. They negotiate hard with their suppliers and show no mercy to their competitors. After all, in business, as in life, only the strong survive.

In America, the best way to acquire alpha
status is to make money – lots of it!

For many, money can become an obsession. Go to any country club, and the favorite sport (after golf) is trying to one-up the other wealthy members. They never mention actual dollar amounts because that would be crass. Instead, they drop not-so-subtle hints about a new Broadway play that they're funding or the wing of the hospital they are sponsoring. They tell you they are being philanthropic, but for the most part, they are still just jockeying for dominance.

But things are different these days; not every leader is an alpha male. A new type of leader has emerged in technology businesses – the omega. Seen by society as nerds, the omega male is technically brilliant but lacks the leadership and social skills of the alpha. They are not inspirational leaders but are typically hard-working introverts who value hard facts over relationships. Despite their intelligence, they have a difficult time understanding the motivations of others. They aren't popular with the opposite sex and only have a few trusted friends. What they do have in common with alphas is that they are ruthless and driven. While they are not killers, they often have the killer instinct.

Bill Gates is a typical omega. Likened to a human calculator, he was not an inspirational leader and never generated the following of Steve Jobs. Considered an introvert, his most trusted advisor turned out to be his wife, Melinda, at least until their divorce. He was, however, a shark in the world of business. It is well known that Bill Gates sold the rights to MS-DOS to IBM without even owning them. Once he closed the deal with IBM, he bought the operating system from a small entrepreneur for pennies on the dollar - without disclosing that he had already sold it. That deal turned Microsoft into one of the most successful businesses in history.

Mark Zuckerberg is another excellent example of a successful omega leader. Like Gates, he is best categorized as a nerd that has proven to be less than an inspirational leader. His lack of empathy and people skills is evident in his continued poor performance in congressional hearings. Like most omegas, he is not a ladies' man. In fact, he created Facebook as a method of rating his female classmates at Harvard after one of them rejected him. Like most omegas, he has a limited circle of trust and married his most trusted friend at Harvard, Pricilla Chan. They have two children, and their bedtime routine involves reading them children's stories and teaching them to code. Ironically, Priscilla won't let her kids go on social media.

Is the transition from alpha to omega leadership another step in the death of evolution?

Indeed, there has been a shift in the status of American workers. In the 1960s, high-status professions included leaders like CEOs, CFOs, marketing executives, doctors, and lawyers. These days, data scientists and software engineers have achieved almost equal status with these professions. The engineers no one would talk to in the 1960s have become sought-after commodities. While they're still not the sexiest, they are considered great catches. They are fabulous marriage material because they make good money, are easily managed, and provide a low-drama environment to raise the kids. If you're lucky, they may even teach the kids to code at bedtime.

Indeed, the concept of management is changing in American business. These days traditional directive leadership has been replaced by collaborative leadership. In this model, no one is superior, and everyone's ideas are, in theory, to be taken at equal weight. In corporate America today, old-style alpha leaders are often considered authoritarian assholes who are now summarily rejected for many leadership positions.

Even the modern interview process is designed to avoid candidates who have alpha leadership traits. In the new model, candidates are given questions designed to express their collaborative and egalitarian nature. This interview style, called the Behavioral Interview, asks questions like "tell me about a time that you worked with a coworker who was not collaborative." In this new world, volunteering in a soup kitchen holds almost as much weight as doubling your market share. Many inspirational alpha leaders are rejected these days in favor of candidates who just get along with everyone.

In the modern business world, just as in evolution, the concept of leadership is dying.

Chapter 9 - Technology

It all started with television. For most Americans, television started the technological revolution that would change our world forever. Sure, there were phonographs, radios, and other forms of technology, but television captured everyone's imagination, and it changed our lives in ways that we're still coming to understand.

If you had a TV in the 1950s, you probably couldn't take your eyes off it. It was so mesmerizing that people would refuse to leave to eat or go to the bathroom. A whole new genre of furniture, the TV table, had to be invented so people could eat dinner right in front of the TV and not miss a minute. The early shows weren't much by today's standards, but it didn't matter because just seeing moving images in your living room was fascinating enough on its own.

By 1960, television was starting to make a substantial difference in the lives of Americans. The broadcast of the first presidential debate between Richard Nixon and John Kennedy was a milestone for democracy. For the first time, voters could actually see the candidates discuss the issues of the day. The debate was pivotal not only in selecting the winner but also in the way the American people viewed politics. Richard Nixon often said that if it were a radio debate, he would have won in a landslide, but his heavy beard and general appearance were no match for Kennedy's good looks and demeanor.

Along with the presidential debate, or perhaps because of it, the evening news became a staple of American television. Instead of reading about events from a newspaper printed the day before, viewers could now get up-to-the-minute information. Rather than spending hours reading the news themselves, Walter Cronkite could read it to them.

In those days, journalism was an honored profession, and journalists were like Jack Webb on Dragnet – "just the facts, ma'am." It was all about who, what, when, where, and why. There was no commentary, talking head analysis, or taking of political sides - just the news. TV journalists and Walter Cronkite, in particular, became the most trusted voices in America. Many of us will never forget the tears streaming down Cronkite's face when he announced the death of President Kennedy.

But over the years, TV became much more than just a way to deliver the news - it became the arbiter of social norms. Since the dawn of man, it was the tribes that set social norms for their members. These standards promoted the continued existence of the tribe because tribes taught the rules for growing up, finding a mate, and bringing kids up properly. Hillary Clinton summarized this ancient truth by saying, "It Takes a Village (to raise a child)."[56] But, with the move to suburbia, people were no longer with their tribes and didn't know the rules. But now, TV was there to help.

TV showed people how to dress, how to talk, how to act, and what to think!

TV showed what beauty looked like with Marilyn Monroe and defined how real men acted with Brando. Shows like Father Knows Best, Ozzie and Harriet, and Leave it to Beaver showed America how to be good parents when they no longer had the tribe to guide them. If you were unsure how to live your life, you could always turn on the TV.

The wonder of TV made us realize that technology, especially electronics, could do almost anything. As a young child, I asked my mother what made the TV work, and her answer was electricity. Then, I followed up by asking her what made the washing machine work, and the answer was electricity. Finally, I asked her what made the lights work, and her answer was electricity. I was so fascinated with everything electricity could do that I eventually got a degree in Electrical/Biomedical engineering.

As compelling as TV was, there was even a newer, even more, captivating technology - computers. The first computers were programmed by decks of hundreds of paper punch cards. There was one computer instruction per card, and if you ever dropped the deck, there was no way to put them back in the proper order. Their primary purpose was to help scientists and engineers solve problems too complex to be tackled manually. Initially, even NASA didn't fully trust these machines and kept a large staff of human "computers" to manually calculate and check everything the electronic computers did.

For many years, computers took up large, air-conditioned rooms in the basements of universities and government facilities. With only one computer per university, anyone granted access to these wonders could consider themselves a pretty big deal. As time went on, computers evolved from building-sized behemoths to pocket-sized smartphones. Rather than being restricted to a few elite technologists, computers are now accessible to almost every human on the planet.

Our smartphones allow us to access all the world's knowledge from anywhere!

America's post-war technology continued to accelerate, allowing us to produce things like supersonic aircraft, moon rockets, and the fiberglass Chevrolet Corvette. It also created inexpensive washers/dryers, vacuum cleaners, and frozen food to help liberate homemakers from their day-to-day chores. This, along with the availability of affordable automobiles, allowed women to enter the business world in large numbers.

Once technology freed females to enter the workforce, it seemed logical that women would continue to pursue parity with men in every aspect of their lives. The women's liberation movement emerged in the late 60s, led by people like Gloria Steinem and Bella Abzug. While the world gives well-deserved credit to these feminist pioneers, it should also be noted that the advent of modern technology also helped many women transcend the life of homemakers.

After thousands of years of male patriarchy – female equality was a pretty scary concept for both men and women. It was so frightening that conservatives like Phyliss Schlafly made an unprecedented effort to stop the equal rights amendment from passing. Portrayed in the fascinating TV series Mrs. America, Phyllis and many conservatives didn't want equal rights to change the traditional roles that women have held for millennia. Amongst other things, they wanted to be sure that women wouldn't be compelled to join the military, enter the workforce, or give up the right to alimony. Apparently, Schlafly was not alone because the Equal Rights Amendment was never ratified.

The transition to the workplace was difficult for many women. Now their time had to be split between bringing up their families and holding a job. A career required strange new skill sets, and females didn't necessarily know the rules. You had to look a certain way, act a certain way, and speak the corporate lingo.

Once they got the hang of it, however, many women achieved great satisfaction from working outside the home.

If you think the women were scared, the men were (and still are) terrified. It was tough enough competing against men. Females were a whole new ball game, and the men didn't know the rules of how to work with them. In some cases, they used sexual intimidation to maintain their workplace superiority. In other cases, they didn't promote women equally or give them equal pay. Nonetheless, women persisted and are now close to achieving parity with men.

Women entering the workforce significantly boosted their family incomes and put a lot of money into circulation. As family incomes went up, the standard of living went up, the money supply went up, and the cost of living went up with it. At some point, economics dictated that the average family could only make it if both the man and the woman worked.

Now both parents have to work to maintain
a good standard of living.

The change in women's roles represented one of the most massive restructurings of social norms in human history - and it significantly changed evolution. No longer was the male the singular alpha breadwinner for the family. No longer was the female the sole nurturer and caregiver of children. For the first time, humans could change the sexual roles that had originally evolved to assure our survival.

In ancient times survival was based on the strength of a person's body, the cleverness of their mind, and the support provided by their tribe. But technology changed all of that. Before modern medicine, many females didn't survive to childbearing age. Even if a woman survived and did get pregnant, many didn't make it through childbirth. Even when healthy children were born, they often died of exposure, starvation, or childhood diseases. It took a robust child with a sound support system to make it to puberty. Once you reached adulthood, your chances were better – but not by much. There were wars, plagues, cholera, dehydration, food poisoning, tetanus, tuberculosis, and countless other threats to your existence. While survival was brutal, it was the way of nature.

Medicine has existed in one form or another since Hippocrates. But scientific medicine came into its own in 1915 with the introduction of Aspirin. Although it is known as a pain reliever, it was the first drug that could save lives by reducing high fevers. The next miracle drug, Penicillin, was introduced in the '40s. It was the first drug able to stop often fatal bacterial infections.

Childhood vaccines started to become common in the 1960s. These vaccines eliminated devastating diseases like smallpox and polio but also helped to manage less deadly diseases like chicken pox, measles, and mumps. In recent times, medical science has also created effective medications for mental illnesses, which, while not fatal, were severely debilitating and often prevented sufferers from finding a suitable mate.

These days practically everyone survives to reproduce.

Medical science continues to enhance survival. Men over 40 are especially prone to heart attacks. Before modern cardiology, many middle-aged men were advised to avoid sex and strenuous exercise and sit quietly in hopes of delaying the inevitable heart attack. Over the years, interventions like stents, anti-cholesterol drugs, better testing, pacemakers, and, ironically, exercise have dramatically extended life expectancy. These days people over 40 are climbing mountains, running marathons, and becoming fathers.

Of course, cardiology is only one success story. Our recent ability to manipulate individual human genes has created a whole new class of genetic remedies, from the mRNA COVID-19 vaccines to permanent cures for blindness. The ability to manipulate individual genes has unlimited potential. Rather than getting your child a nose job at 16, you could splice out that "big nose" gene shortly after birth to have them naturally grow the perfect nose. In a more Orwellian case, wealthy parents could have their children infused with genes to make them great athletes or have high IQs.

Manipulating individual genes can
stop evolution in its tracks!

But enhancing survivability is just one aspect of how modern medical technology has impacted evolution. The second rule of evolution is to reproduce. But, on May 11, 1960, technology broke that rule when the birth control pill was approved for use in the United States.[57] The introduction of "the pill" had a profound effect because it allowed women to decide when and if they wanted to have children. This became a tremendous liberator for females and a significant enabler of women's move into the workforce.

The people who used the pill, especially at the beginning, were the planners. These upwardly mobile people think about how much kids cost to raise, the effect on their careers, and if they will have enough time to bring them up properly. These folks are the doctors, lawyers, engineers, accountants, programmers, managers, CEOs, and entrepreneurs of the world. These people also have highly developed prefrontal lobes that help them to defer immediate pleasure to achieve a greater goal.

The pill was a godsend - but it may have accelerated the death spiral of evolution!

These planners are precisely the people that evolution should promote, yet with modern birth control, they are the people reproducing the least. No matter where you live, the higher your education and income level, the fewer children you have.[58]

In a May 2013 Huffington Post Article entitled People Getting Dumber? Human Intelligence Has Declined Since the Victorian Era, Dr. Jan te Nijenhuis of the University of Amsterdam, points to the fact that "women of high intelligence tend to have fewer children than women of lower intelligence. This negative association between I.Q. and fertility has been demonstrated time and again in research over the last century."[59]

We can now control practically every aspect of reproduction. We can stop fertilization with birth control or enhance fertility through in vitro fertilization. We can bring together sperm and eggs outside the body and inject the resulting embryo into a uterus – anyone's uterus.

These days, we can create the baby of our dreams by selecting DNA from the "right" donors. The singer Melissa Etheridge has been "out" as a lesbian for quite some time. When she wanted a child, she picked singer David Crosby from Crosby, Stills, and Nash as the sperm donor. I bet that kid can sing!

These days almost anyone can make a baby!

But technology hasn't just changed the ability to reproduce; it has also affected how we are attracted to potential mates. Make-up has been used for thousands of years to help attract mates. The ancient Egyptians, especially Cleopatra, were noted for their advanced makeup 5000 years ago. Perfumes, another attractant for the opposite sex, have been around for at least as long. But, in the last 50 years, technology has taken us well beyond makeup. Plastic surgeons can now completely reshape the human body. Breast enhancements were one of the earliest and most obvious ways to increase a woman's desirability. In recent years other enhancements, like butt implants, liposuction, lip injections, nose jobs, and various other procedures, have become enormously popular.

Cosmetic enhancements, at least in theory, give women a higher propensity to reproduce. These procedures help women attract top-tier males when their natural appearance would have constrained them to mate at a lower level. The problem is that cosmetic surgery doesn't change your DNA - so what you see is not what you get. We hope the parents aren't disappointed when the baby somehow doesn't look like the mother.

The popularity of Kim Kardashian as a standard bearer of modern beauty shows to what ridiculous lengths this has gone. With a body shape not found in nature, Kim has unapologetically enhanced almost every part of her anatomy to produce a hypersexualized caricature of a real woman – and became a billionaire doing it.

Cosmetic surgery fakes out evolution.

Of course, in the case of the Kardashians, Caitlyn Jenner probably holds the family record for cosmetic enhancement. While we celebrate the ability of technology to let Caitlin lead her best life, you do have to wonder what this does to the concept of evolution. While gratifying for the individuals, we are again intervening with our species' ability to select appropriate biological mates.

It could be argued that one thing that differentiates humans from all other species is the ability to create technology. Sure, some animals use rudimentary tools they find in the environment, but only humans conceive, design, and build ever-advancing technologies.

For most of human history, these technologies enhanced human's ability to observe the first rule of evolution – survival. Tools, weapons, motorized transportation, and later technologies like antibiotics all enhanced humanity's ability to survive. But, starting in the '60s, technologies like the birth control pill and in vitro fertilization dramatically changed the nature of the second law of evolution – reproduction.

As much as technology changes reproduction, it changes evolution on an even more basic level – environmental adaptation. As you remember, evolution is based on an organism's ability to survive changes in its environment. Giant leaps in evolution typically occur when the environment changes, making it more difficult for a species to survive. If they can't adapt, they die, as evidenced by the hundreds of species that go extinct every year. If they survive, the DNA that enabled that survival gets passed on to future generations.

But our technology makes humans the only animals capable of changing their environment. We live in buildings that control the temperature within a few degrees. Clean water can be obtained within seconds. Food is cold stored in your kitchen and warmed instantly by microwaves. Technology can control most diseases, and we have weapons to control all our natural predators (except each other). With our level of technology, we can survive almost any external event except perhaps an asteroid strike or nuclear war. So, there is no reason for biological evolution to move forward.

Once the technology genie is out of the bottle, there is no going back because technology dramatically improves individual lives. The question is, what happens to humanity as a whole?

With technology's influence on our ability to survive, reproduce and our environment, evolution has no mechanism to move forward.

Chapter 10 - The Media

It was the best of times. The year was 1963, John Kennedy was Commander in Chief, and all was right with the world. But Lee Harvey Oswald changed all of that. I remember walking home from school that day and seeing all the distraught moms on their front porches, waiting for their kids to come home. I'll never forget the forlorn expressions on their faces, their red eyes, and the tears streaming down their cheeks. They looked as if they had lost one of their own children.

While the great debate between Kennedy and Nixon demonstrated the power of television to inform. The slow-motion scenes of the bullet impacting Kennedy's skull demonstrated television's ability to horrify. The televised image of Kennedy's three-year-old son, John Jr., saluting the flag-draped coffin was a tear-filled goodbye to a time that might have been.

I wondered in later years if we would have been as upset had Nixon, Johnson, or Bush been assassinated, and the answer was clearly no. Kennedy was a leader of vision, compassion, and intelligence. He was the ideal alpha – the man we wanted to follow to our destiny. We lost more than our president that day; we lost our dreams.

And JFK was not the last dream we would lose. Martin Luther King, the man destined to lead the black community to the promised land, was murdered in 1968. The assassination of John Kennedy's brother Bobby only two months later obliterated any hope that the legacy of the Kennedy era would continue beyond his death.

Kennedy's assassination marked an inflection point in the history of the world.

Our country, and to a large extent the world, was left rudderless, adrift. The unfortunate combination of the Vietnam War and the weak leadership of Lyndon Johnson created divisions in the country. The women's liberation and civil rights movements added to that turbulence. To make matters worse, young people were not only demonstrating against the war but starting to subscribe to a new movement – the hippies.

This historic turmoil was covered every night in excruciating detail on prime-time TV. While past wars were only publicized through newspapers or highly edited newsreels, Vietnam was shown in nearly real-time. Graphic scenes of napalm, live fire, and body bags we're beamed into people's living rooms nightly. On the few nights, Vietnam didn't lead the coverage; it was civil rights marches, college anti-war protests, or the hippies in San Francisco.

The hippie movement was about a lot more than headbands and bell-bottom jeans. It was a movement that rejected the traditional values of family, capitalism, and American exceptionalism. It was, in a sense, the antithesis of the post-WWII "Happy Days" culture of the 50s and 60s.

The hippie's favorite motto was "don't trust anybody over 30", and their favorite pastimes were sex, drugs, and rock and roll.[60] They rejected materialism, capitalism, and especially the Vietnam war. Although it is now viewed as one of the most extraordinary cultural events of our time, the Woodstock music festival was viewed with horror by "the establishment" (not to mention the parents of the kids who attended). The world was shocked at news footage of naked drug-crazed teenagers having sex in the mud and listening to music loud enough to make them deaf. The world was clearly out of control!

For the parents of the hippie generation, the middle-aged Americans who fought in WW II, the world had gone insane. During World War II serving in the military was an honor. Most people volunteered to go. In the case of Vietnam, young people were running to Canada to avoid being drafted.

Watching the 6 o'clock news led to thousands of uncomfortable dinner-time conversations. Pro-Vietnam World War II-era parents on one side of the table arguing with their hippie-inspired kids on the other side. This conflict was immortalized by the classic show All in the Family. In the TV shows of the

'60s, like Fathers Knows Best and Leave it to Beaver, the father was portrayed as an all-knowing alpha male who could fix any problem. In contrast, All in the Family represented the head of the household, Archie Bunker, as a bigoted, ineffective ignoramus.

For millennia, children had learned about the world from their tribe. Generation after generation passed these beliefs to their children through a rich legacy of storytelling and mythology designed to keep the people in the tribe in line. These were the sacred values that helped the tribe survive.

In the early 20th century, immigrants like the Jews, the Italians, and the Irish identified themselves as members of a tribe. In those days, you could ask anyone on a busy Brooklyn street what they were. Rather than say American, they would immediately respond with their ethnic identity, "I'm Italian" or "I'm Irish." Until radio and television became available, you generally had the same worldview as your tribal community. But the post-World War II generation was brought up on television, and now the tribes were being dispersed from the ghettoes into suburbia.

The dispersal of the tribes had immense implications that no one could have predicted. The tribe morphed into the ethnic immigrant communities that not only provided a common belief system but an ideal social support system. Having an extended family, neighbors and friends who lived in the neighborhood meant that someone was always looking out for you. You had aunts, uncles, cousins, and friends to rely on when things were tough – and those people were all there to ensure you succeeded. The movie, My Big Fat Greek Wedding was all about the tribe - a big proud family constantly extolling the virtues of being Greek.

It would be a mark on the community as a whole if one of their own did something that would embarrass the community. Whether you were Italian, Irish, Greek, or Jewish, no one wanted to see a member of the tribe get in trouble. Whenever a member of the tribe was in the news for a crime, the older people would shake their heads from embarrassment and say something like, "did it have to be one of us?".

But when the tribe dissipated, people lost much of their social support. Without the tribe, people not only lost role models but also their community's love and support. These days, many people not only don't interact with their neighbors, they don't even know who they are. Without a tribe - it's easy to become isolated, lonely, and angry. Does this contribute to the depression, crime, and mass violence that occupies most of the evening news?

With the dissipation of tribes, there was no
one to take care of marginal individuals.

Tribes took care of their own regardless of whether you had a physical disability, a behavioral issue, or just didn't have a job. If you needed help, they provided it - whether it was someone to push your wheelchair, someone to lend you money, or just getting you a job with uncle Leo. Tribes would be embarrassed to let their members be homeless or panhandling in the streets. But when the tribes broke up and moved to the suburbs, they lost that proximity that made the tribe work. Could the break-up of these close-knit communities set the stage for the epidemic of homelessness we see in the current era?

During the Vietnam war, for the first time in history, children didn't want to fight on the tribal team. They didn't want to dress and act like their parents. They didn't want to listen to the same music that the tribe listened to. They didn't want to accept women's subservient roles and didn't want to believe that other tribes, like black Americans, were inherently inferior.

Modern media encouraged people to break away from their tribal values.

This quiet revolution moved into every aspect of life. People began to question the traditional sexual morals of their tribe. They didn't want to be fixed up by matchmakers and lose their virginity on their wedding night. Instead, they wanted to shop in a veritable candy store of beautiful people with whom you could have sex almost at will.

The United States' stunning defeat in Vietnam eroded the sense of national unity that World War II had created. It also massively eroded our image as Superman around the world. No longer were we the invincible country that stood for Truth, Justice, and the American way. But worse than looking bad in the eyes of the world, we started to look bad in our own eyes. We began to lose confidence as a nation.

Ronald Reagan was just the right person to bring America back together after the Vietnam War. He was a soothing communicator and genuinely interested in making America great again. Regan changed the focus from Vietnam to the real enemy at the time - Russia. Without firing a shot, he presided over the fall of the Soviet Union. Reagan's policies, such as the uber-expensive Star Wars missile system, helped to destabilize the Soviet government to the point where it fell. The United States was once again the world's alpha dog.

But all of that changed on September 11th, 2001, when terrorists attacked the World Trade Center in New York. This was the worst trauma the United States had ever suffered in the homeland. It was far worse than Pearl Harbor because it attacked New York City, the symbolic center of everything that represents America! The country hadn't felt so vulnerable since the attack on the mainland in the war of 1812.

That morning, the United States went from the world's most confident and powerful nation to a nation in shock. It wasn't just Americans who cried that day. People across the globe shed tears because they knew there was no more Superman to save them. Those hijacked aircraft were kryptonite, and it's safe to say that the United States will never be quite the same.

While we blamed the attack on Osama Bin Laden, we knew that the mighty USA could be attacked by any one of hundreds of low-tech terrorist cells. It took nearly ten years to find Osama bin Laden. The fact that the mighty United States took a decade to catch a man hiding in holes in the ground further showcased America's new impotence.

As bad as 9/11 was, one thing made it worse – the internet. The web gave everyday people the ideal tool to disseminate 9/11 conspiracy theories. In the past, conspiracy theories were typically spread from person to person, whispered by word of mouth. Before the web, if someone didn't believe a conspiracy theory, or if it sounded crazy, they would not pass it on to anyone else.

Now that everyone had access to the web, every conspiracy theory and piece of misinformation was passed on to thousands of people. Misinformation blew up like an atom bomb. Just as splitting atoms would, in turn, split more atoms producing an uncontrollable chain reaction, the internet enabled misinformation to "go viral." Like a nuclear chain reaction, once misinformation was posted on the web, it was unstoppable.

Tons of conspiracy theories swirled, including the idea that 9/11 was an "inside job" to justify the American war on terror. Another 9/11 conspiracy theory purportedly came from a missile loader at Andrews Air Force Base who was there when the jets scrambled to intercept the hijacked plane over Pennsylvania. The official story is that the passengers tried to take the plane back from the terrorists causing it to crash into an empty field. But according to this conspiracy theory, one of the scrambled jets returned to base - one missile short.

While email and computer bulletin boards were available as early as the late 90s, Facebook and Twitter made social media ubiquitous in the early 2000s. Unlike network news and newspapers, no one fact-checked social media. There were no confirming sources, no editor, no threat of being sued for libel - nothing. Anybody could say anything and not only get away with it - but generate millions of followers in the process.

The introduction of the iPhone on June 29, 2007, exponentially increased the influence of the internet. It was like going from a hand grenade to an atomic bomb. People now had social media in their pockets 24/7. While social media was already available on desktop computers, the ability to check your phone out anytime and anywhere was a complete game changer.

We all knew people who couldn't take their eyes off their phones for more than a minute. You would go to dinner, and they plop their phone on the table and keep at least one eye on it throughout the entire conversation. Every few minutes, their face would take on a serious expression as they ignored you, picked up the phone, and began to type on it with furious intensity. From the look on their faces, you thought they were responding to a medical emergency or liquidating their stock portfolio.

Then, with a sigh of emotional relief, they would put the phone down and turn their attention back to you. You would think the crisis was over but eventually realize that your friend would only engage with you for a few minutes until the next tweet came in. This became even more disappointing when you asked what was so important, and it turned out to be someone's selfie or doggie picture. We called these people phone junkies.

It turns out that we were right. The ability to continuously interact with your phone anytime and anywhere made it literally addicting. Scientific studies have concluded that constantly looking at social media (especially getting likes) lights up the same pleasure centers in the brain that drugs do.[61] [62]

Social media can be as addictive as drugs.

A 2020 paper entitled Neurotransmitter Dopamine and its Role in Social Media Addiction published in The Journal of Neurology and Neurophysiology, amongst others, concluded that positive reinforcement on social media releases a flood of pleasure-producing dopamine into the pleasure receptors of the brain.[63] Once those pleasure centers are repeatedly stimulated, the brain craves even more stimulation whenever the dopamine begins to dissipate – just like a heroin addict.

Dopamine produces euphoria in the pleasure centers and a feeling of reward for accomplishment.[64] For social media addicts, each post provides a hit of dopamine leads to a feeling of satisfaction. In other words, dopamine gives people the sense that they have accomplished finding the "correct" answer and that more research isn't necessary.

> *Could dopamine release be why people so stubbornly cling to information they receive on social media – regardless of accuracy?*

About the same time the iPhone was introduced, Fox News hit the airwaves. Although it billed itself as a news outlet, it was created for a single purpose – to make money for Rupert Murdoch. Its strategy was to tap into the right-wing, male, middle-aged audience turned off by the left-leaning newscasts of the big three networks.

Murdoch hired Roger Ailes, a former variety show producer and political consultant, to design Fox News more as a form of entertainment than a legitimate news channel. Choosing hosts more for their appearance than their journalistic credentials, he trotted out a collection of sexy blondes like Megyn Kelly, Gretchen Carlson, and Laura Ingraham to titillate the right-wing white male audience.

> *The formula worked - Fox News became America's most popular news network.*

While Fox News (or CNN) wasn't as addictive as social media, there was still an element of addiction. The fact that the same stories were repeated continuously, 24 hours a day, could still produce dopamine and stimulate the pleasure centers – especially if it was a position that you felt strongly about. The more stimulation the pleasure centers receive, the more they crave.

Probably nothing points to the merger of news and drama better than the OJ Simpson incident. Everyone was riveted watching OJ drive that white Bronco in the slow-motion getaway attempt from his wife's murder. Here was real life in real-time. America was held spellbound not only on the night of the murder but for years to come.

While murders are committed every day, this was the ideal TV drama. "The Juice" was the NFL running back who had transcended the race barrier with his infectious smile, enormous talent, and stunning white wife. He was the embodiment of the American dream, not just for black folks, but for everybody. Never had a pop hero fallen so far – so fast.

OJ wasn't the first American hero to fall, nor the last. Over the subsequent years, it seemed like the majority of our heroes were shown to be fatally flawed. Drug addiction, infidelity, financial impropriety, domestic violence, and racism were just some of the issues that seemed to affect practically everyone that you looked up to. It afflicted everyone from Lance Armstrong who was doping for most of his career, to Hollywood stars like Charlie Sheen, who spent more time in rehab than in his dressing room.

And let's not forget about Bill Clinton. Clinton seemed to be the golden boy of American politics with his meteoric rise to the presidency. His charisma, intellect, and hair was as close to John Kennedy as any politician had come. But his highly publicized affair with intern Monica Lewinski ended both his career and tenure as a role model.

There were no more American heroes
- including America itself.

The loss of our role models had many implications. The most obvious one is that there was no one to look up to. Without good role models, there was almost no one to pattern our lives after. Role models are essential to the species' survival because they set an example for what's expected.

Studies have shown that about 90% of people are followers.[65] They are supposed to follow their bosses, politicians, priests, and generals. Then there are the even smaller percentage who inspire us, like astronauts, actors, and football heroes. Either way, humans evolved to use role models to teach them successful life strategies.

Those mirror neurons that help children learn by imitating the behavior of those around them continue to function throughout our lives. You easily observe them in action when you walk into a new situation, like a new job or a party, where you unconsciously mimic the body language and speech patterns of those around you.

Just observing good (or bad) role models causes the mirror neurons to automatically and unconsciously cause us to emulate them.

But in the last twenty years, something strange happened - something entirely against the conventional laws of evolution. Having flaws went from a liability to an asset. Led by a bunch of public relations fixers, a new narrative emerged. You can act any way you want as long as you claim you were a victim. Whether it was breaking the law, lying, cheating, or addiction - you were absolved from responsibility if you apologized and got therapy.

Shows like Jerry Springer, Maury Povich, and Geraldo Rivera sensationalized the flaws of real-life people. The popularity of these shows, along with the emergence of reality TV, meant that TV (and the internet) displayed far more instances of "bad behavior" than "good behavior." The existence of mirror neurons made this bad behavior contagious.

Our role models went from good to bad,
and naturally, we imitated them.

Adding to the problem was the fact that these TV shows did not place any judgment on their participants. Never married with six babies from six different fathers – no problem. Cheating on your husband with his brother – why not? These shows made the phrase "don't judge me" wildly popular. Although community judgment was one of the essential tools that tribes used to maintain their culture - it was now out of style.

Admitting you had flaws was not only accepted; it became all the rage. Now people weren't judged on how good they were; they were judged on what they had to overcome. It might have been drug abuse, bad parents, ADHD, or other hardships. People with healthy bodies, intact minds, and an ethic of hard work were an archaic vestige of times past.

This new paradigm even made its way to college applications. Essays about how you were the captain of the football team that won a state championship were rejected in favor of applications detailing how you overcame some obstacle. Perhaps no one in your family had attended college before, or you overcame dyslexia to achieve a B average. Whatever the story, college applicants learned very quickly to exploit their weaknesses as well as their strengths to achieve their goals.

Our culture gradually changed from one based on facts to one based on compelling stories. The media learned that facts don't sell advertising and don't hold eyeballs because they don't get people emotionally involved. Old-time newscasts like Walter Cronkite were actually kind of boring. The "just the facts" approach to journalism probably put more people to sleep in front of that TV than the glycogen loading from those big old meat and potatoes dinners that we used to eat.

People love a good story. It's been that way since tribal leaders told stories in front of their campfires. Even back then, a good myth took precedence over fact. Once people like Roger Ailes, Mark

Burnett (producer of The Apprentice and Survivor), and Jerry Springer got ahold of this concept, there was no turning back.

With our heroes gone, and a world full of highly flawed role models, it's hard to see how society can evolve forward.

Chapter 11 - Generations

It's said that children are our future, but in the year 2023, we're not so sure. Today's kids are clearly living differently than they were in the happy days of the 60s. You can blame it on a lack of traditional values, over-parenting, under-parenting, bad schools, TV, the Internet, the lack of role models, or COVID, and you would be right. But these disturbances might also be caused by something more profound, like a radical shift in our evolutionary model.

Watching kids grow up can be an incredibly beautiful thing! All parents dream their kids will make friends, build a career, find a supportive mate, and produce wonderful grandchildren. But nowadays, it's tough even getting them to move out of the house.

In times past, young Homo Sapiens were encouraged to go out and explore life. Starting as children, the tribe and their parents prepared them for life's challenges. The idea was to toughen them up for the cruelties of the real world. In many tribes, this culminated in a coming-of-age ceremony.[66] Around puberty, boys would be sent into the forest to kill an animal or battle an enemy to prove they could survive in the real world. Those who didn't have the skills to pass the test often never made it home. Accordingly, their DNA was never passed on to future generations.

*Coming-of-age ceremonies helped eliminate
DNA that didn't promote survival.*

These days, young people do most of their exploring in another dangerous world – a virtual one. While the digital world is generally less harmful to the body, it has wreaked tremendous havoc on the emotional well-being of our kids.

Young people used to be adventurous. They were convinced that they were invincible and, therefore, often a little reckless. But a little recklessness is necessary if we are to push the species forward. Christopher Columbus was reckless when he braved uncharted waters to find the new world. The Wright brothers were reckless when they launched that first airplane into the skies. And the Apollo astronauts were reckless when they landed the lunar module on the moon with their fuel running low. We will never accomplish anything meaningful if we are always playing it safe.

Christopher Columbus, navigated by the stars, to a mythical new world. Today, people can't even find their way to the grocery store without using Google Maps.

As digital technology becomes ubiquitous, the skill level necessary to perform even the most common tasks has been minimized. People can instantly Google anything from How to Get a Promotion to How to Survive in the Wilderness. If you have a question, Siri, Alexa, ChatGPT or the many "experts" on social media are there to answer it – no thinking is required. These days kids consider playing Call of Duty without enough virtual ammo packs a high-risk activity.

Our world no longer promotes the risk-taking and
critical thinking skills we need to survive.

Worse yet, the digital world doesn't allow for the physical side of human interaction. Lost is the eye contact, the tossing of hair, the knowing smile, the sideways glance, and the proud or submissive body postures. Lost are cues like watching someone sweat, their face flush, or their slight nervous tremor. The loss of physical connection with the world and the warmth of a human touch can be devastating. The combination of fear of the physical world and lack of experience with genuine human interaction is bad for everyone and impedes our ability to evolve.

In the 60s, children played blissfully throughout their neighborhood. Parents knew the neighbors would gladly watch out for them. Now fearful parents keep their children sequestered inside. If for some reason, they are allowed outside, they are helmeted 100% of the time. Why take a risk when in the end, everyone gets a trophy anyway.

When kids skinned their knees in the 60s, their mom threw a Band-Aid on it, so the kid would not drip blood on the floor and sent them back outside. These days, many parents would have rushed that same kid to urgent care.

When I was four, I somehow understood that once you started school, you wouldn't be able to stop. The prospect of giving up my beloved cartoons and, worse yet, having homework was repulsive. So, when the school bus arrived that first day, I started kicking and screaming. After a minute or two, the bus driver and my dad grabbed me, picked me up, and carried me onto the bus (once I got there, I decided that I liked it). Fast forward to 2022, when a friend told me her child made a similar fuss about starting school. Rather than forcing him to get on the bus, she took him to therapy for six weeks until the therapist said it was okay for him to start school.

Today, young people are taught that the world is a dangerous place. In my day, we couldn't wait to go to the DMV on our 16th birthday to get our licenses. That day marked our independence – the day we were free to go wherever we wanted, whenever we wanted. But these days, many kids don't even ask to get their driver's license when they turn 16. They are afraid to drive, and we suspect, to some extent, afraid of the real world.

This fear might also be a subliminal reason that more and more young adults are choosing to live with their parents for an extended time. According to the Pew Research center, more young people live at home than at any time since the Great Depression. According to Pew, only 29% of young people lived at home in 1960, as opposed to 47% in 2020.[67] Since COVID, of course, the numbers have skyrocketed even further.

> *Young people are delaying leaving the*
> *nest and finding a mate.*

That said, the world is indeed more dangerous than it was in the 60s. In addition to rampant street crime, mass shootings occur almost every day. Mass shootings, especially school shootings, are among the most despicable crimes imaginable. While there is no possible justification for these crimes, they can at least be somewhat understood by our evolutionary model of the six types of males.

We have discussed alpha and beta males, but mass shootings are almost always performed by a category known as delta males. The opposite of an alpha, delta males are the loners and misfits of society. Sometimes they were born that way, but often they were converted to deltas by a traumatic event. It might be not making the team, being rejected by a female, or getting fired. Deltas are angry, and that anger builds as they begin to suspect that they are destined to be one of life's losers. They believe their life has not amounted to anything because other people have not treated them fairly. Once this image takes hold, he unconsciously reinforces this belief by sabotaging himself at every opportunity. Deltas are known, for example, to find many ways to blow job opportunities, from not showing up to having angry outbursts during interviews.

In tribal days, the tribe would ensure that these individuals had social interaction and a productive place in the community. But these days, without a tribal support system, the deltas, who are loners to begin with, are almost completely isolated. Many take refuge in the virtual world where, unlike in the real world, they have control. They often play video shooting games where the gun is the power source. This creates a strong association with a gun as a source of accomplishment and desensitizes them to the horrendous damage that an actual shooting produces.

Most mass shooters fall into two different age groups. One group is middle-aged and most likely going through a midlife crisis. At this point, they finally realize that they have almost no status in society and have accomplished very little with their lives. The other group is the 18 to 24-year-olds who are starting to realize that even though they made it through high school, they will likely never amount to anything. The younger group is more likely to be involved in school shootings because they may be angry about the school experience or that kids in school rejected them. In either case, shooting innocent people allows them to vent their unbearable anger and gives them a once-in-a-lifetime opportunity to know what it feels like to be a powerful individual.

In the mind of a mass shooter, you can't be a powerful alpha without a powerful weapon - and that weapon is the AR-15. This military-style assault rifle is the choice of practically all mass shooters because it can quickly inflict tremendous damage. Handguns are essentially designed for defense. If you want to maximize carnage and go out in a Rambo-like blaze of glory, the AR-15 is your gun.

Mass shootings are fueled to some extent by the media. Many, if not most, mass shooters allude to the possibility that they will commit an attack on social media weeks, if not months, before they go through with it. It may be a cry for help, but more likely, it is an attempt to build their base of followers so that when the event does occur, someone is watching. This effect is boosted by extensive TV coverage of these horrendous crimes that give perpetrators a chance for their 15 minutes of fame.

In many cases, these mass shootings result in the perpetrator's death, and in most cases, it is by design. Rather than just taking their own lives, many opt for "suicide by cop" as a final act of defiance. To some extent, this may be just another manifestation of the ever-growing rate of teen suicide that is turbocharged by social media.

> *Social media amplifies every event and emotion – especially for young people.*

But, even beyond the influence of social media, young people seem to be more fragile than at any time in history. Take the case of the USS George Washington. In 2022, seven sailors assigned to the USS George Washington committed suicide.[68] This was not during combat – but while the ship was in drydock in Virginia.

And don't forget the drugs. We're not just talking about the crack cocaine that devastated the black community in the 80s but opioids that are currently killing approximately 70,000 people from every socioeconomic class each year.[69] Like Spice from that science fiction classic Dune, our society seems to need drugs. Whether it starts as opioids for back pain, Xanax for anxiety, or recreational cocaine, people seem unable to face everyday life on their own. It's no wonder that parents never rest easy.

These days kids are fragile and parents are scared.

This general anxiety and confusion carry over to every aspect of young people's lives, including one of the most essential activities - sex. The sexual revolution of the 60s took sex from something to be ashamed of to an activity to enjoy. Everyone wanted to look sexy, act sexy, and be sexy. After all, the ability to attract a mate and procreate is one of the most critical components of evolution.

Interestingly, the gender neutrality movement seeks to eliminate all traces of biological sexuality. Instead, it promotes the narrative that people can choose their sex and that the sex assigned at birth is irrelevant. Rather than emphasizing their sexuality, which has been the evolutionary imperative since the dawn of humanity, gender-neutral people downplay their sexuality to the point that it doesn't exist. The movement has generated such popularity that many businesses require workers to put "their pronouns" in their signature block to reduce the chance of offending someone.

Think about the last time you saw a genuinely well-dressed man or woman on the street. When you did, it probably turned your head. Today, everyone is in T-shirts and athleisure. While it may be comfortable, it doesn't always present your most attractive self – and attractiveness is an essential precursor for mating and reproduction.

If attractiveness wasn't important, why would the peacock's plume have evolved?

Speaking of attractiveness, Americans are becoming obese at an alarming rate. According to the Centers for Disease Control, middle-aged men are 27 pounds heavier and middle-aged women are 25 pounds heavier than they were in 1960, roughly a 16% increase.[70] The prevalence of diabetes has gone up 10 fold from less than 1% in 1958 to 11% in 2022.[71]

In the late 60s, the media promoted skinny models like Twiggy as an ideal for female beauty. This unrealistic ultra-slender standard of beauty caused some young people to develop anorexia and bulimia. But America is now being taught by the media to embrace obesity as positive and healthy. Walk into any Target or Walmart, and you will see posters and mannequins depicting people who are medically obese.

The motivation is simple – if you tell people that they look beautiful as they are, they are more likely to identify with your products. TV spots, magazines, and digital ads that promote average women's bodies have a far larger audience and make more money. The problem is that the average American's body isn't healthy.

It's as if people don't want to be attractive anymore!

Sports Illustrated is usually a showcase of the world's most athletic people. But in 2016, it featured 201-pound model Ashley Graham on the cover of the swimsuit edition. Our intent is not to fat-shame anyone or deny that being too thin can cause issues. However, sending the message to young women that it is okay to be clinically obese isn't good either.

Many say that weight is subjective and that many paintings from past centuries portray plump people as the proper standard of beauty. But only the well-fed lords and ladies could afford to commission those portraits. Back then, being a zoftig was a status symbol. It showed the world that you were wealthy enough to overindulge.

There is, in fact, one medically ideal body type regardless of current fashion. That body type has a low Body Mass Index (BMI) with a high percentage of muscle and a relatively low percentage of fat.[72] Despite the media's attempt to make obesity beautiful, clinical obesity is extremely unhealthy and certainly doesn't promote the survival of our species. Ask any doctor!

In some circles, losing weight is now considered a bad thing. A Time Magazine article published on June 17th, 2022, entitled

Forget Physique, Mental Health is the Newest, Hottest Fitness Goal implies that worrying about physical appearance is detrimental to mental health.

Consider the infamous 2019 Peloton Christmas advertising campaign. It portrayed a super-fit model who profusely thanks her husband for buying her a Peloton bike for the holiday. Rather than celebrating fitness, the social media haters were all over the ad because they imagined the man was "forcing her to exercise" and he was buying her a Peloton because he "didn't find her natural body attractive enough."[73]

The October 2018 cover of Cosmopolitan featured 279-pound model Tess Holiday in a swimsuit with the cover headline "Is Success an Illness." An October 5, 2022, Time magazine article entitled Ambition is Out echoes a similar theme.[74] These shocking headlines make sense only in a society where everyone gets a trophy for being just the way they are. Pushing hard to achieve a goal is apparently now considered obsessive, abusive, or both.

This, of course, is in stark contrast to what evolution promotes. From the earliest days, the ability to attract a mate and produce children was essential in the tribe. Ancient tribe members used make-up and tattoos to enhance their sexual attractiveness. Things are different now. In a nod to gender neutrality, many women are reducing or eliminating make-up. Tattoos, traditionally a male art form, are now at least as prevalent among young women as young men.

*Are people getting tattoos to be a badass or as
an unconscious nod to tribal traditions?*

Ancient tribal people knew the importance of attracting a mate. And while make-up and tattoos helped, they had a secret weapon – dancing. Dancing evolved as more than just a fun time; it is a powerful way to attract a mate. Hundreds of animal species utilize elaborate mating dances to attract mates. Watching the grace and athleticism of potential mates allows animals to unconsciously determine if the dancer is a suitable genetic partner. Watching the moves evokes hormones from the midbrain that can cause arousal in both men and women.

Dancing has always been important. It was prominent in the 1920s big band era, progressed through the USO swing days of the 40s, and into the "sock hop" days of the 50s. But sexually suggestive dancing peaked in the 70s and lasted through the disco era.

The dances of that time were designed to display body gyrations that might be taken as a preview of moves in the bedroom. The popular movies Saturday Night Fever and Dirty Dancing were made to pay tribute to that sexy dancing. The tiny miniskirts that the girls wore made the dances even more enticing, and the new technology of breast implants rounded out the look. During the disco era, no one made any bones about seducing someone with their dancing and taking them home. The Donna Summer classic Last Dance details what they hoped for back then, a partner, at least for the night.

But that was then. Now young people simply jump up and down on the dance floor. Rather than grinding with someone of the opposite sex – they dance with everyone simultaneously, male or female, in a massive undulating crowd. Today's dance moves no longer seem to be forms of either art or seduction.

Even sexy movies don't seem to appeal to young people anymore. There has always been an element of sex in the movies whether it was the innuendo of the 1940's or the full-frontal sexuality of movies like Basic Instinct in the 90s. In a 2003 Time magazine article entitled Why Aren't Movies Sexy Anymore author Eliana Dockterman posits that sexy movies are no longer appealing to young moviegoers.[75]

Today's most popular young adult movies come from the Marvel comic books. While Superman always had Lois Lane, most of today's Marvel superheroes don't have sexual relationships and don't get the girl in the end. Make no mistake; those Marvel superheroes are seen as role models for many young people, as evidenced by the massive attendance at Comicon and the sold-out Avengers movie openings.

Even in the height of their reproductive years, young people seem to be afraid of overt sexuality. They seem too shy to date outside of groups, and even established couples often invite friends along on dates. They include friends and family in what used to be the most intimate and romantic moments of courtship. Young males in high school are expected to perform an extravagant public "promposal" to get a date for the prom. Asking someone to marry you is now a choreographed affair with a photographer and sometimes dozens of people hiding in the bushes to witness the event and congratulate the happy couple. Rather than a romantic honeymoon of lovemaking in a paradise like Hawaii, people arrange destination weddings so they can party it up with their friends.

Young people prefer to meet on social media rather than in real life. So rather than being attracted to someone by how they dance, the lilt in their voice, or how they flip their hair, attraction is now measured by how cleverly they text.

Studies have confirmed that people are attracted not only by appearance but the way potential partners smell.[76] This should be no surprise, considering how other mammals, like dogs, sniff around to select a mate. People use the word pheromones to describe the sexually attractive scents secreted to attract potential mates. How someone smells can provide a good indicator of their biochemistry, including their hormonal content. So far, however, no dating app produces pheromones.

No longer bound to their tribal taboos, mixed-race and inter-religious relationships are at an all-time high. That, along with gender fluidity, has given young people an almost infinite number of choices - and yet, finding a mate seems more challenging than ever. Like Kate Beckinsale dating Pete Davidson, the matches preferred by evolution no longer seem to apply.

If all of this wasn't enough, the #MeToo movement came along. The #MeToo movement was long overdue and will stand as one of the great inflection points in gender equality. While it went a long way toward addressing the horrible injustices that many females have endured - it scared the shit out of men.

Many men are now afraid to buy a girl a drink because they don't want to be accused of spiking it with date rape drugs. Some men are reluctant to make eye contact with strange women and unwilling to flirt for fear of being considered a creep. They definitely won't kiss anyone without explicitly asking for permission. We are told that before things proceed any further, asking for explicit consent on video is not unusual.

Then came COVID. Kids were already increasingly living at home with their parents – which never helped anyone's sex life. But COVID meant they were home for years with no prospect of having a physical relationship with anyone.

Many people in this century will never find a mate.

It's no wonder that dogs are now so popular with singles. Go to any apartment building where young people live, and you'll see a steady stream of single people leaving the building to take their dogs for a walk. A dog is a good way of meeting people of the opposite sex. But, more importantly, dogs help mitigate loneliness.

These days, dogs are treated almost as well as people's children and may, in many cases, be surrogates for them. These once mighty predators have been so domesticated that behaviorally they bear almost no resemblance to their ancient wolf ancestors. Like humans, survival of the fittest no longer applies. Many dogs are so fragile that they now require anti-anxiety medications or hug-simulating garments to keep them from freaking out when they are left alone.

Evolution depends on sexual attractiveness and mating, and both are in trouble. It's hard to say how much sexual fluidity, remote dating, and sexual anxiety contribute to the precipitous birth rate drop in first-world countries. The annual birth rate in the US has gone from 22.6 per thousand in 1960 to 12.0 in 2022.[77] In 2020, there were only 1.3 births per woman in the United States, far below the replacement level of 2.1.[78]

This can't be good for the survival of the species.

Chapter 12 - Politics and Religion

Ninety percent of us are followers.[79] It's not our fault - it's how we are wired. Like the wolves that follow that alpha leader to find prey, humans have found that following a leader is better for survival than acting independently. For hundreds of thousands of years, the propensity to follow has been hormonally, neurologically, and psychologically programmed into our brains.

Leaders and followers have different hormonal make-ups.[80] Various studies have shown that alphas have relatively high levels of testosterone (strength, fearlessness, posturing, fighting), serotonin (focus, calmness, stability), and low vasopressin (aggressiveness).[81] Conversely, followers measure higher in cortisol, the stress hormone which makes them panicky in stressful situations like confrontations or battles. In other words, followers are generally not biologically that comfortable with leadership.

Having an alpha leader reduces the stress of followers by lowering their cortisol and increasing oxytocin, the love hormone. We see this with dogs, the once mighty wolves who are now so domesticated that they have intense anxiety when left alone. The minute their human alpha returns, the dog feels a rush of oxytocin that creates feelings of trust, attraction, and love. That is why dogs are so happy when you arrive home!

Like dogs, people don't have much conscious control over whom they trust and to whom they are attracted. So, it doesn't matter whether the leader is trustworthy or knows what they're doing; once people get hormonally engaged, they will follow the leader anywhere. Because the leader/follower relationship is hormonal, people can become extremely passionate about following a particular leader or cause.

Unwavering loyalty is very useful for
leading warriors into battle!

Human leadership has far more profound implications than the leader of a wolf pack. This is because the leaders of nations can control the destinies of millions of citizens. As tribes grew into civilizations, leadership encompassed all aspects of the tribe's survival. A complex society has many moving parts, and good leadership is essential if that society is to survive. Ancient leadership determined what crops would be planted, where to build the cities, how to obtain water, and when to move the tribe if food became scarce. Modern leadership decides when to unleash the nukes.

But, unlike dogs, human leadership is about more than just hormones – it is also wired into the brain's prefrontal lobe.

Various MRI studies have found that good leaders have highly developed frontal lobes, specifically the right frontal lobe.[82] As it turns out, this is the source of not only future planning but what we call vision. Vision has been the hallmark of our most inspirational leaders, from Alexander the Great to Martin Luther King.

So, for the 90 percent of us whose frontal lobes are not quite as developed, we need a leader. The leader provides the vision along with the hormonal chutzpah to make it happen. The vision isn't always good, but the alpha leader is hormonally and neurologically endowed to inspire people to follow them.

Leadership comes in many forms. It could be a tribal leader, the quarterback, a political leader, a spiritual leader, or a business leader. As long as the group can identify with the message, any talented alpha can evoke the same oxytocin-fueled sense of passion, loyalty, and love.

But, these hormones are not just produced in response to the leader; they are also generated in response to being a member of "the team." While having a leader reduces your cortisol levels being part of a team drops that stress hormone level even more. The team can be any group: Americans, Republicans, your company, your faith, or your favorite sports team. Watching an NFL game on any given Sunday demonstrates how team identification can evoke powerful hormone-driven emotions, not just for the players but for the fans!

*People have a biological urge to be part of
something bigger than themselves!*

Religions are tribes that provide another example of how the
human brain responds to leadership. All religions have one
thing in common: they follow the ultimate alpha leader - God.
It doesn't matter which God you believe in or whether he or she
actually exists. What does matter is that the belief in God makes
followers feel good.

Interestingly, people's brains can also generate these hormones
without needing a physical leader - simply by believing. Brain
studies show that belief in God increases oxytocin and decreases
cortisol, just like following a human leader does.[83] We follow
human leaders because we think they have a vision and a plan,
but who could possibly have a grander plan than God? For
individuals, belief in God is a win-win situation. It comforts
them and gives them hope, and it's not subject to the possibility
of human error.

The concept of faith may actually be based on neurobiology. The
loose definition of faith is a belief that defies logic. For people of
faith, the hormones evoked by their midbrain can suppress the
logical prefrontal lobe to allow for the existence of something
greater than themselves.

While religions generally reinforce "good" behavior, a cult is
different. A cult is a group that follows a charismatic living
leader who often asks followers to do things that are counter

to their own good or the good of society. Over time, the leader can build such power over the members that they will follow any command regardless of the repercussions. In 1969, Charles Manson convinced his cult "family" to murder five people in Hollywood. In 1978, Jim Jones had so much control over his followers that he convinced all 909 of them to commit suicide by drinking poison "cool aid."[84]

This level of control requires the constant repetition of beliefs as an emotional hook. Constant repetition of these beliefs or even thinking about these beliefs produces a flood of "love" and "calming" hormones that activates the brain's pleasure centers. The more this activity is repeated, the more the pleasure centers expect and crave the stimulation. As a result, the follower increasingly seeks to listen to the leader, read the scriptures, or pray. When the midbrain's pleasure centers are stimulated often enough, it becomes a positive feedback loop that can't easily be escaped from.

People in cults describe it as a feeling of overwhelming love.

In extremist religions or cults, repeated hormonal stimulation hijacks the pre-frontal lobe connection, blocking the ability to think logically. You may have experienced this effect yourself during times of extreme stress like an emergency. During these times, high levels of hormones like adrenaline and cortisol can invoke an extreme fight or flight response, causing panic and blocking your ability to think logically. Similarly, very high levels of love and pleasure hormones like dopamine, norepinephrine, and oxytocin can also stop logical thought. We experience this when we are in love or in a cult.

Terrorist groups like ISIS use this to their advantage. These groups are very much like the forces commanded by medieval warlords – and their followers are willing to hurl themselves into almost certain death for the glory of their leader.

Charismatic leaders keep the terror group members in a constant state of hormonal flux, alternating between the fear/stress hormones and the stimulation of pleasure centers to provide relief. This technique was depicted in the show Homeland, where a captured US prisoner is turned from an American patriot into an embedded terrorist by a charismatic leader through hormonal cycling.

After capture Brody is subjected to merciless torture. At regular intervals, he is brought to see the terrorist leader, who treats him with love, bathes his wounds, and offers him food. After repeating this sequence many times, Brody eventually becomes addicted to the pleasure hormones evoked by sight and the sound of the leader. He becomes irrevocably brainwashed to do the leader's bidding, even when it involves killing hundreds of Americans.

It is natural to think of middle eastern terrorist groups as tribes because they fit most of the traditional descriptions of an ancient tribe. But there are lots of less formal tribes in our modern world. Religions, businesses, labor unions, and even our political parties function as tribes.

Think about the Democrats and Republicans as two competing tribes that coexist on the same land. In that context, it's easy to understand why they don't get along. Each party has its sacred tribal values, and like most tribes, the members adopt those core beliefs in their entirety. After all, publicly deviating from the tribal core belief system traditionally meant ex-communication.

Republicans tend to believe those who work hard deserve to hold on to everything they have earned. They favor laissez-faire capitalism, constitutionalism, low government control, preserving individual freedoms, low taxes, and states' rights. Most Republicans buy into their entire agenda even though it is not totally consistent with their philosophy. Rarely do you hear Republicans state publicly that they are pro-abortion even though banning abortion removes some of the personal freedom they hold dear.

Democrats tend to believe everyone should be supported and provided with a "decent life" regardless of their circumstances or personal achievements. They believe that government needs to intervene to offer services to everyone, especially those less fortunate. Their sacred belief package includes immigration, abortion rights, unions, limiting big business profits, criminal justice reform, the environment, and gun control. Like the

Republicans or any other tribe, they generally adopt the complete package of sacred values. For example, you rarely see Democrats who say that the best way to help the poor is to help big businesses flourish so they can provide more jobs.

*If you want to stay in the tribe, you
must stick with the party line.*

Whether you are a Democrat or a Republican, towing the party line has many advantages. The first advantage is that you're thinking has been done for you. There's no sense stressing over the day's issues when your party leadership tells you exactly how to think. Another advantage is that your brain generates the hormones of love and belonging from membership in the group. The same feelings of warmth you get when you're on a team can be generated by being part of a political party through the secretion of oxytocin. The third reason is that deviating from the party line increases stress hormones and risks rejection from the larger group. In the end, group affiliation goes back to our basic premise that most people are followers.

Because membership in a political party produces hormones, political discussions can evoke very passionate feelings. Of course, the same could be said of religion. That's why the old adage says "never discuss politics and religion."

Interestingly, the two political parties philosophically fall on opposite sides of evolution. The Republican Party, at its core, believes in a Darwinian survival of the fittest. They think that the people who are the fittest physically or intellectually should be the ones that rise to the top. Republicans believe that underprivileged people are better served by allowing evolution to select rather than giving them handouts which ultimately hurt the process of natural selection. They think that things like labor unions prevent exceptional performance and that welfare programs give people an excuse not to work.

Republicans believe in free market capitalism – the survival of the fittest philosophy of commerce. This is an extension of the ancient you eat what you kill. They believe you will be successful if you are intelligent, work hard, and follow the rules. Sharing it through taxes that become handouts to less fit individuals destroys the work ethic - which hurts everyone.

Democrats, conversely, have a philosophy that is, to some extent, anti-Darwinian. They believe that regardless of their physical or mental traits, everyone is equal and deserves an identical piece of the pie. In the democratic model, people don't prosper because mostly they don't have the initial advantages - and it is the government's job to compensate.

Democrats like Elizabeth Warren believe that big business is inherently corrupt and that they are stealing money that belongs to the workers. They support unions as an equalizing force that helps employees have the power to oppose the alpha managers.

They believe corporations should share their wealth with the workers who have helped create it.

When discussing our political leaders, you often hear people say, "Are these the best we can come up with?" The problem is that most elected officials are alphas; and being a servant of the people or subservient to anyone is antithetical to an alpha's nature. In the real world, there are two qualifications to be an elected politician 1) having enough ego to think you can pull it off and 2) being able to convince others to follow you. Both are fundamental alpha traits.

Despite our perceptions, our leaders aren't necessarily wise or altruistic. They are ambitious alphas who seek power - and that power isn't necessarily used to improve the country. Senators, in particular, epitomize the old-school alpha male with their thick manes of hair, expensive suits, and fiery rhetoric. They claim to be doing the will of the people, but alphas do whatever they want.

We think politicians are elected on the issues, but
most people just vote for the dominant alpha.

While our politicians are elected to improve the country, there is little motivation for them to actually do so. There are a couple of reasons for this. First, most politicians base their campaigns on catering to a small base – not the majority. The base is usually passionate because they are strongly polarized to the left or the right - and only passionate people contribute big bucks. When candidates get elected, they take it as a mandate to continue with these polarized ideas even though most of their constituents are closer to the middle.

Doing the will of the people doesn't always play well within the party's larger strategy. Joe Manchin was one of the only Democrats opposing several bills before congress in 2021. Whether he thought he was representing his constituency or making a name for himself, he was often the only one keeping the democratic strategy from being implemented. The pressure on him was enormous. It is much easier to go along with the party than to represent your constituents.

Rather than determining leaders by who was the most-deadly warrior or who had a crown passed down to them from their ancestors, democratic leaders are chosen by the people they rule. But, while an improvement, this model still favors the election of alphas over other, perhaps more qualified betas, gammas, or omegas, who may be better at crafting policy. As we discussed in the early chapters, alphas are the kind of leaders that people want to follow. It doesn't matter if they are smart or have extensive experience; it just matters that people perceive them as leaders.

The Obama administration was full of highly competent, well-educated people who, for the most part, had the country's best interests in mind. But despite Obama's incredible speaking ability, he and his administration were perceived as nothing more than policy wonks - people who were engineering the country rather than leading it. They didn't present themselves as alpha leaders.

Obama's Secretary of State was Hillary Clinton. Few would doubt Hillary's intelligence, academic credentials, or experience. Hillary may have been one of the most qualified people ever to run for president. But just like with Al Gore before her, the American people didn't want a brainiac; they wanted an alpha leader.

Donald Trump, on the other hand, was the ultimate alpha male. He made his reputation by being a fearless negotiator reminiscent of the old Rockefeller-style tycoons that first made America great in the early 1900s. It was no wonder the Make America Great Again theme resonated with voters. His image was reinforced weekly for 15 seasons of The Apprentice, where "the Donald" was portrayed as an alpha king making the weekly pronouncement "you're fired" - the business equivalent of "off with your head." Donald even led the lifestyle of a king. He had "castles" all over the country, all decorated with the same color scheme – gold. He even named his son Barron!

In many ways, Donald Trump was a
throwback to the monarchy

He was bigger than life, brash, opinionated, and didn't really care what anyone else thought. His TV personality served him well during the Republican debates, where he clearly overpowered would-be alphas like Ted Cruz.

Like any true alpha, he had women – all beautiful. Regular guys don't marry international models who look like Melania. Even his daughter Ivanka was beautiful and intelligent. If she wasn't the product of superior DNA, then who was? Okay, Eric didn't qualify, but Donald was smart enough to keep him mostly out of sight.

And then there was the hair. That crown of blonde hair swirled like a combination of "Surfer Bro" and bad comb-over. A crown of thick hair is an alpha-bestowing hallmark of many of our politicians. Most male US Senators have a thick mane of hair – like a lion. John Kennedy had great hair, as did Ronald Regan and Bill Clinton. The crazy golden hair of British Prime Minister Boris Johnson or Donald Trump may subliminally remind voters of the golden crown worn by monarchs.

When Trump hit the stage for the Republican debates, he probably knew less than anyone else standing on that stage. But he used the techniques he needed to intimidate the other candidates. Trump made up derogatory names for his opponents, gave them degrading looks, and used the word "small" a lot. When he debated Hillary, he stalked her on the stage. While most of us thought this was creepy, it did, in a weird way, establish his dominance. He wasn't the best-behaved candidate on that stage, but good behavior is for betas. Alphas do whatever the hell they want. So, with no experience in governing, Donald Trump defeated Hillary Clinton, one of the most experienced politicians in the modern era.

But Donald's ultimate secret weapon was Twitter. During the campaign, he tweeted multiple times each day. Every time his followers read his tweets, their brains would produce instant shots of oxytocin and dopamine, feel-good hormones that stimulated their pleasure centers and made them feel that the message was true. The content of the message was irrelevant because the dopamine reduced the desire to analyze the statement. During his campaign, his followers would get approximately 15 tweets per day, and each tweet would produce a rush of hormones. During his presidency, Donald Trump posted 26,237 tweets.[85]

*When you stimulate someone's pleasure center
many times a day - they get addicted!*

Just like a smoker or a drug addict, overstimulation of the pleasure centers disables the mechanism of logical thought. They will do anything to get their fix, including rationalizing a lot of really bad behavior. So, in many ways, hardcore Trump supporters actually became addicted to the cult of "The Donald." As Trump himself said in 2016, "I could stand in the middle of 5th Avenue and shoot somebody and not lose any supporters."

People will disagree on whether he was a good or bad president - but we can all agree that he didn't want to leave office. True alphas like Trump hate to lose. Losing hurts their self-image, and for a guy like Donald Trump, who made a career of calling everyone else losers, the thought of actually being a loser was unbearable.

So, rather than having to consider himself a loser, Donald Trump made up The Big Lie. Hitler's propaganda minister, Joseph Goebbels, said, "If you tell a lie big enough and repeat it often enough, people will eventually come to believe it, and you will even come to believe it yourself".[86]

Trump wanted to believe he won - the way a kid who knows better wants to believe in Santa Clause. Otherwise, he would have to admit to being a loser. When trusted members of his administration told him that Joe Biden had legitimately won, Trump looked for others who would agree with him.

Even when the courts found no evidence of election fraud, he looked for independent parties who would support his fantasy. And on January 6th, Trump addressed thousands of people in the mall who did support his story. How could he resist having them march on the capital to show that he was a winner after all?

Chapter 13 – Civil Unrest

We never dreamed it could happen in America – but it did. On January 6th, 2021, an army of American citizens stormed the Capitol building in an attempt to overturn a free and fair election. And an army they were; dressed in military fatigues and Kevlar vests, the mob descended upon the capital like a human tidal wave. Armed with reconnaissance reports and battle plans, their mission was to prevent the constitutionally mandated transfer of power - even if they had to hang Mike Pence to do it.

Right-wing militias like the Proud Boys had been preparing for something like this ever since Donald Trump publicly told them to "stand down and stand by" in September 2020. Ironically, these are the same groups that consider themselves some of the staunchest believers in the Constitution.

America is arguably the most successful country that has ever existed. Even during COVID, the US economy was the best in the world. Times were so good that many people quit their jobs to stay home and live off the government's largess. So why, of all the countries in the world, was there an insurrection in America?

Many people have attributed America's success to the fact that it is a "melting pot" of people from every race, religion, and country on earth. We were a nation where immigrants were proud to take on a new identity as an American.

That model worked early in the 20th century when the Irish, Italians, Jews, and other ethnic groups made their way to our shores. The immigrants worked hard and ensured their children looked and acted as if they were born here. They prayed that when their children went to school, they would be considered 100% American by their peers - and they dreamed of sending them to college. Coming to America was, after all, a dream come true.

For most of the 20th century, the dream of a diverse America living, working, and playing together seemed to be moving forward – albeit slowly. It reached its high point in the '60s with Martin Luther King and new legislation designed to improve the situation of minorities. But make no mistake; we were still a nation of tribes.

When Facebook was introduced in 2005, there were high hopes that social media would be the thing that finally brought us together.[87] Finally, people from different backgrounds could meet and exchange ideas through technology. Many thought we would break the shackles of our tribal boundaries and become a highly evolved society where everyone worked together - singing Kumbaya.

While it's true that social media enabled communication between people of all types, it didn't necessarily promote constructive communication. It disproportionately gave loud voices to people with fringe opinions. While the majority stayed silent, social media enabled fiercely vocal haters to unleash their radical ideas on the public. Rather than a forum for understanding, the web became a place where people could actively select only the information sources which agreed with their preexisting position. It became an echo chamber.

Instead of people from all walks of life coming together to exchange ideas, social media became an amplifier for people at the extremes. Rather than helping our society evolve past tribalism, social media promoted the formation and fortification of new tribes. These tribes were not based on common DNA but on radical ideas. And these new tribes had their own sets of sacred values.

Social Media was a nice dream – but it backfired.

Social media did indeed bring people together who would have never ordinarily met. But much of the time it coalesced people at the fringes together to achieve critical mass. This mass created a chain reaction that allowed homegrown radical groups to explode like an atomic bomb with new members and even more radical positions. To make matters worse, hackers from Russia, China, and other countries also began to use social media to destabilize the country.

In 2014, a Russian troll farm called the Internet Research Agency began to target American voters with propaganda and polarizing messaging.[88] At the same time, Russian military intelligence (GRU) would distribute fake news through made-up personas and phony think tanks.

The Russian GRU also hacked the Democratic National Committee and leaked that sensitive information to organizations such as WikiLeaks. Those leaks became one of the biggest side stories of the 2016 election. But, it turned out that the GRU's main mission was to help Donald Trump get elected. Once the leaks were exposed, social media outlets started to crack down on fake accounts, necessitating a strategy shift for the Russians. Rather than using false content to influence voters, they realized they could have more leverage by polarizing online conversations - helping to amplify our existing divisions.

The increasing division began to peak as the 2016 election approached. Donald Trump, of course, used Twitter to keep his supporters at a fever pitch. Adding to the social media onslaught was highly partisan coverage by outlets such as Fox News and CNN. The commercial popularity of Fox News spawned new conservative outlets like One American News (OAN) to capture viewers who thought that Fox News was too liberal. But the divisions created by social media didn't stop after Trump was elected - if anything, they got worse. Throughout the Trump administration, the country continued to become more polarized.

Social media was the most significant influence
in the election of Donald Trump.

Then COVID happened. Now, instead of just glancing at the news and social media, everyone was stuck at home with nothing to do. The TV and internet were on all day. Everyone was desperate for information, especially about the pandemic. And, if you wanted up-to-the-minute news, you weren't watching ABC, CBS, or NBC; you were watching either CNN or Fox. Not only did Fox News and CNN viewership go up, but the number of viewing hours went through the roof. While people were watching the COVID updates, they were also being exposed to many stories that were not COVID related. Of course, when people weren't watching the TV news, they were on social media. The propaganda disseminated by both the right and the left was causing everyday Americans to choose a side.

During the pandemic, my family often played a game we called The Real Story. It entailed watching a story on CNN and then turning to Fox to see how they covered it. In many cases, the same story was hardly recognizable or wasn't covered at all. The opposing rhetoric of the two news organizations drove an even bigger wedge between the left and the right.

COVID was like pouring gasoline on a fire. Rather than being the common threat like World War II that brought Americans together, it divided us even further. Simple public health measures like wearing a mask became highly controversial. Opposition to the mRNA vaccine, a genuine scientific breakthrough funded by Donald Trump, ironically became the focal point for Trump supporters. Even the kindly doctor Anthony Fauci received death threats from the far right.

Despite government assurances that COVID-19 was a mutation of the naturally occurring SARS virus, the internet was abuzz with conspiracy theories. Most claimed that China had created the virus in a Wuhan lab and was distributing it worldwide to bring down the west. Some white supremacists theorized that the "superior" DNA of the American people rather than the vaccine allowed America to handle COVID better than China.

Conspiracy theories aside, many people refused to take the vaccine as more of a protest against government intervention than real scientific concerns. Early in the pandemic, many police officers, firefighters, teachers, and others were willing to lose their jobs (and perhaps their lives) rather than comply with vaccine mandates. While some suspicion of a brand-new vaccine could be understood, the anti-vax movement gained more steam after millions had already taken the vaccine without serious side effects. Numerous televised hospital bed pleas from dying anti-vaxers begging people to take the vaccine seemed to make some people even more determined not to get it.

The anti-vax movement was a convincing demonstration of the power of fake news. When anti-vaxers were asked why they wouldn't take the vaccine, you would hear an almost verbatim set of lines from their favorite anti-vax blog. Standard retorts included, "I'm still doing my research," "it gives you heart problems," or "natural immunity is better" (even though you have to suffer through COVID to acquire natural immunity). But it was more than misinformation. It signaled that these folks didn't trust the federal government and didn't want to be told what to do by "the man."

But beyond the conspiracies, it also reiterated the far-right's entrenched belief in survival of the fittest. While many right-wing extremists don't believe in evolution per se, they had faith that the strong would survive. While they didn't say it outright, they believed that their tribe's superior DNA would protect them from COVID. They also didn't mind if COVID eliminated the weaker members of the herd.

Although he started, funded, and took credit for the vaccine, Donald Trump saw the direction of his base and backed away from any association with the vaccine.[89] He also decided to move away from recommending masks. If he wanted to stay in the good graces of his base, he had to disavow all government-mandated COVID protocols. After all, the election was coming up.

Trump learned that the bigger the lies he told about COVID, the more popular he became. So, when the election results rolled in and it became apparent that Joe Biden had won, Trump had no problem telling The Big Lie. He knew that, like with COVID, his base didn't trust any government institutions. Trump knew that the more government officials verified that Joe Biden won the election, the more the far right would believe that the election was a conspiracy.

January 6th was the ultimate manifestation
of government distrust.

Although Trump eventually left the White House, he did leave one crucial legacy - his Supreme Court. For the first time in decades, the Trump-appointed Supreme Court was dominated by highly conservative right-wing justices. While most of these appointees were known for their legal scholarship, they were put there for a singular purpose – to further the conservative agenda. Trump knew these justices would put their right-wing tribal values above their intellectual logic. Despite our high expectations for judges, they listened to their sacred values.

Let's not fool ourselves; liberal justices decide cases in line with their sacred values, and conservative justices do the same - regardless of the legal arguments. After they choose, they pull out the law books to rationalize their decision with convoluted legal justifications that put the Jewish Talmud to shame. Their impressive prefrontal lobes are often used simply to justify the answer their "hearts" have known for years.

This particular court consistently makes decisions that reduce the federal government's power and increase states' rights. The famous reversal of Roe versus Wade was more about whether the federal government could ban abortions nationally than abortion itself. The 2022 case concerning concealed carry for weapons in New York State was about whether individual states have any right to impede constitutional gun rights. The court also ruled that the Environmental Protection Agency, the federal agency established to control air pollution, did not have the right to enforce air pollution mandates. The court wanted individual states to enforce pollution standards even though it is impossible to confine smog to a given state.

And it's not just the Supreme Court promoting states' rights. During COVID, people in Republican states would not enforce federal guidelines for masks and vaccine mandates. Red states were also reluctant to abide by the federal voting guidelines that created so much acrimony during the 2020 election. The Republican states seemed to have had it with federal government control.

We haven't been this divided since the Civil War.

Worse than the distrust of the government is our distrust of each other. Discussing politics has now become a dangerous pastime. Almost all of us have lost friends because they were firmly on the other side of the political divide. Rather than discussing our issues, we are like two boxers, each standing in our corners – willing to fight it out when the bell rings but mostly trying to dodge and weave around political questions. Like the Sharks and the Jets in West Side Story, we have learned to hate each other – and are more willing to take each other out than talk it out.

In an August 2022 CBS news poll, 49 percent of Republicans called Democrats "enemies" instead of just political opponents.[90]

The same survey also found that 64 percent of Americans think political violence will increase. This many Americans have rarely agreed on anything.

If we continue down this path, we will devolve into 50 states that all do things differently. The United States will lose all of the efficiencies it gains by running national programs in the environment, transportation, commerce, and other areas. With the European Union coming together to exploit the synergies of doing things together and China taking an even more highly centralized approach, we are bound to lose our place as the leader of the world economy.

There is a lot more at stake here than bragging rights. The U.S. is currently the world's only Reserve Currency. That means that the value of all of the world's currency is judged by its value compared with the dollar. This also means that we can effectively print all of the money we want without effectively devaluing the dollar.

But if we lose our economic dominance, there will be pressure from the EU and China to also use their currencies as the world's Reserve Currency. That could be a very slippery slope because that could limit our ability to print money to "juice" the economy. It also means that people from other nations will be more comfortable buying bonds from those other nations. Both of these things will reduce the funds flowing into the U.S. and cause the economy to shrink.

With states making their own rules, it will be complicated to control climate change. Many states will eviscerate any current carbon-limiting legislation, and their greenhouse gases will spill into neighboring states.

With no federal control and the Colorado River running dry, water disputes between the fast-growing states of California, Arizona, and Nevada are front and center. California desperately needs water to continue its role as a leading producer of crops, while Arizona wants to add a million new homes to the Scottsdale area in the next decade. In the February 1st, 2023, Washington Post, Joshua Partlow wrote an article entitled As the Colorado River Dries Up, States Can't Agree on Saving Water where he states, "Two decades of drought have pushed critical reservoirs to dangerously low levels."[91]

If this disagreement over water rights continues, will one state or another send its National Guard to protect its water infrastructure? Will this result in armed conflicts? Wars have been fought for less.

Some states will start to ban things like telemedicine to be sure that abortion pills are not prescribed over the internet. They may ban gender-affirming surgeries, hysterectomies, and even vasectomies. At some point, state police will probably try to block a shipment of birth control devices going into their state. This may require federal troops to escort it into the state the way black students were accompanied to the University of Alabama during the George Wallace era.

In March of 2022, the Washington Post Magazine published an article entitled "They are Preparing for War."[92] The article interviewed professor Barbara Walter of the University of California, who specializes in studying Civil Wars. She concludes that the United States is "coming dangerously close to the conditions required for Civil War." While we think of Civil Wars as something from the past, don't forget the terrible civil conflicts in Northern Ireland, Rwanda, and, more recently, Myanmar. Like our Civil War, these conflicts featured ordinary citizens killing each other over "tribal" differences.

Instead of being divided between north and south along the Mason-Dixon line, we will now be divided east and west into red and blue states. The coasts will be primarily Democratic and more liberal, while the center and the south will be principally Republican. And this isn't just a division of political parties; it is a division of sacred values.

Different cultures in each state may lead to violence.

Even today, many people would like to move to a state more aligned with their political tribe. Conservatives are already leaving their jobs in east or west coast cities for states more aligned with their values like Texas and Arizona. Liberals will be inclined move to blue states where they can access abortions and feel their voting rights will be upheld.

And the dividing line is not just about political philosophy. It's also racial. Well-educated Republicans can often work from home and get jobs wherever they want. Less well-educated minorities often have blue-collar jobs that require them to be on-site. These jobs are more prevalent in the larger cities, primarily on the east and west coasts. Over time, the red states will become redder, and the blue states will become bluer.

Currently minority populations are growing significantly faster that the white population, and with the abolishment of abortion in certain states, that gap will accelerate. That shift could bring the Democrats back into power. Still, with a relatively young Supreme Court, it will take at least another generation to reconsider many of these states' rights issues.

Many have predicted that race wars may become inevitable. The evening news regularly features videos of police killing black men for minor infractions. The beating of Trye Nichols in January 2023 by black police officers is an extreme example of the brutality that law enforcement regularly brings to bear.[93] The result is the erosion of public trust in yet another government authority. It also results in additional rules that will further limit the role of the police.

After the death of Freddy Grey on April 19, 2015, in Baltimore, the Baltimore police force has been limited in their powers by consent decrees and the fear of igniting civil disobedience.[94] After that incident, they also found it extremely difficult to recruit new officers and have remained chronically understaffed. The result is that the Baltimore police department has become highly ineffective, and crime has skyrocketed. Of course, many cities are experiencing the same problems.

Civil disobedience will undoubtedly increase as our differences are heightened, and the police are weakened. It may be racial, political, or criminal, but at some point, the police will be no match.

There will also likely be disputes over illegal immigrants. Recently Texas has been unilaterally shipping migrants to New York City and Washington, DC. At some point, states may deploy law enforcement to keep immigrants from being transferred to other states. Texas will likely want to take immigration into its own hands and create some sort of state border patrol – and someone will get hurt.

Texas, and perhaps other states, may try to secede from the union. While it may not become an armed conflict, the case will likely wind up in the Supreme Court to further define states' rights. At some point, states may question whether their residents must pay federal income tax.

Emboldened by the insurrection, expect more violence from right-wing militias. They will march on state capitols, bomb legislators' homes, and probably even hang a few people. These groups will be joined by many less radical right-wingers and survivalists who have been preparing for a civil war for a long time. While they don't often refer to it as civil war, almost all of them will tell you, "When something happens, we will be ready."

Right-wing militia groups will increasingly take matters into their own hands, and it is not likely that red state governors will be willing to pay the political price incurred by stopping them. Of course, organized crime will also take advantage, and street gangs will rise in prominence. We could quickly devolve into states controlled by warlords like Mexico.

We could be fighting each other for a long time.

Currently, there are 393 million guns in the United States, more than one gun per resident.[95] Obviously, this is far more than needed for hunting and defending their homes from a home invasion. Even more disturbing is the fact that there are currently over 20 million AR-15 assault weapons in the hands of the American public.[96] These AR-15s are not appropriate for hunting or self-defense; They are strictly offensive weapons of war.

Many Americans are ready for civil unrest - why else would they stockpile millions of weapons?

Chapter 14 - Artificial Intelligence

Evolution has ended. The combination of modern agriculture, electricity, industrialization, infrastructure, biotechnology, the Internet and now artificial intelligence means that our individual survival characteristics are no longer needed for our species to survive.

Being a ferocious warrior is no longer a requirement to survive. These days, wars are fought with technology rather than brute strength. Today's most lethal soldier isn't a muscular brute, but a nerdy drone pilot who stayed home playing video games on prom night.

In the 21st century, you don't have to be the best hunter to feed your family. With supermarkets in every strip mall and fast-food restaurants on practically every corner, you not only don't have to know how to hunt - you don't even have to know how to cook.

Modern humans don't have to learn to find water, make a fire, find shelter, and sew animal hides together to survive the winter. You can hibernate in your heated home, have food delivered by Grub Hub, and binge Netflix until spring arrives.

Our world is changing far faster than we can evolve!

Indeed, today's best provider may be a finance graduate from Wharton or a coder working for Google. These days, you don't have to recruit a pack of warriors to change the world; you do it by recruiting followers on social media. Your ability to write snappy text on dating sites may better predict your ability to find a mate than your physical appearance.

In times past, leaders, heroes, and stars were created because of some genetically evolved talents like athletic prowess, bravery, or good looks. But these days, you can buy a cosmetically enhanced body, and heroes are created more by publicists than genetically derived traits. While Superman could be only controlled by Kryptonite, the TV show The Boys shows that future superheroes can be controlled only by their corporate public relations department.

The only thing you need to survive these days is money!

In 2013, Eric Topol coined the term for the type of person ideally suited for life in the 21st century - Homo Digitus.[97] As opposed to Homo Sapiens, who evolved to survive in the physical world, Homo Digitus are human beings who evolved to handle the digital world. While Eric Topol's vision of Homo Digitus evokes images of humans evolving long fingers, giant brains, and superhuman intelligence, the truth is quite the opposite.

Modern technology has removed practically all of the pressure for humans to evolve in any particular direction.

Think about today's world. We are sequestered in climate-controlled environments. The sun is so intense that our skin quickly burns without protection. Rather than sitting around the campfire telling inspirational tales of our sacred truths, we're sitting around the TV learning our social norms from reality shows. We communicate with each other through digital media and find that even talking on the phone is becoming obsolete. Live, personal interaction is now by appointment only.

We are bipolar – we have animal minds but live in the digital world.

We live schizophrenic lives where our animal brains tell us one thing, and the modern world tells us something different. Our ancient midbrains still want to follow the alpha male. But brute-like alpha behavior is becoming unacceptable in our modern world. What we used to think of as strong leaders are now considered by many to be authoritarian asses.

Men's animal brains still want to stare at a pretty girl, but society has taught them that they better not be seen doing it. Females who used to fantasize about testosterone-filled bad boys now see well-behaved beta software engineers as ideal mates. Not only do they bring home the bacon, but they are much easier to manage.

Modern women bring home their own bacon and can have fulfilling lives with or without a mate. Procreation is no longer women's sole reason for being, and many are torn between the joys of motherhood and the pressure to break the corporate glass ceiling.

Our leaders are now made through the media rather than their physical prowess. And, while Donald Trump did express all the old-time alpha characteristics, he would never have been elected without social media technology. Today's thought leaders have generated millions of followers not through stirring rhetoric but through 280-character tweets on Twitter. They're called influencers for a reason, and many have sowed the seeds of division within our society.

Our "bipolar" existence is manifesting
through rifts in our society

In a May 22, 2013, Huffington Post article entitled "People Getting Dumber: Human Intelligence has Declined Since the Victorian Era," the authors state that "Westerners have lost 14 IQ points since the Victorian era.[98] That translates to us losing approximately 1% of our intelligence every decade. One of the contributors to the article, Dr. Gerald Crabtree of Stanford University, said, "The reduction in human intelligence would have begun at the time that genetic selection became more relaxed. I projected this occurred as our ancestors started to live in more supportive higher density societies".

*The trend towards lower IQs is accelerating
as fast as our technology.*

Digital technology has made everything easier for everyone. Thousands of software engineers spend their entire day thinking of ways to make websites and apps intuitive enough that everyone can instantly use them. Instruction manuals are now a thing of the past (except for Ikea). These days everything from bank statements to school testing is designed for the lowest common denominator. As a result, any advantages that the "smart kids" might have had are rapidly eroding.

That erosion is now accelerating exponentially with the advent of Artificial Intelligence (AI). While digital technology is great at multiplying human abilities, AI is taking machines beyond leveraging human ideas to creating original ideas.

AI has shown that it can do many jobs better than most humans. With just a few prompts, art packages like Jasper and Midjourney can let novices produce art that competes with Salvadore Dali. Chat GPT can write an original poem that sounds just like Edgar Allen Poe. School kids have discovered that AI is even smarter than the kids who used to let them copy their answers during exams! The trend has gone so far that ChatGPT recently released a tool to allow educators to detect AI-generated papers to help put the brakes on this trend.

The fact that educators need a special tool to determine if a human or a machine produced a paper means that we are very close to having a machine pass the legendary Turing Test. The Turing Test was proposed in 1950 by the seminal computer scientist Alan Turing to answer the question, "Can a machine think?"

The Turing Test entails a blind text conversation between two humans and a computer. If the machine can successfully fool the people to the point that they believe it is human, then it has passed the Turing Test. According to Turing, a device must actually be thinking if it can be clever enough to fool humans during a random conversation. While many dispute the validity of this test, no other definitive test has taken its place. So far, no machine has officially passed the Turing Test, but with the current state of AI, it will likely be passed by the time you read this.

Once machines pass the Turing Test, many people (and almost certainly the law) will consider them thinking, sentient beings. Based on animal rights legislation and other legal precedents, this means that these machines will have rights.

Consider the case where one sibling wants to terminate the AI avatar of its late mother, but the other sibling wants it to exist forever to act as a virtual grandmother for the grandchildren. To keep the avatar "alive", the second sibling goes to court, presenting the argument that the avatar is a conscious, thinking being as proven by the Turing Test. The argument continues from that point to show that except for a biological body, the

AI avatar possesses the characteristics of a human and that terminating it would amount to murder. Any judge would be hard-pressed to argue otherwise.

Will we ever be allowed to terminate the existence of a conscious AI avatar?

Once it is demonstrated that AI is enough like a human, it may be guaranteed every right that humans have. The legal and moral dilemmas will be almost infinite. Will AI be allowed to vote? Will they be allowed to marry? How will they be able to have their own money? Will they own property? Can they run a company? Will they collect alimony after breaking off a long-standing relationship?

There is already a tremendous fear of AI, and the fact that we are starting to realize that AI might be granted all of the rights of humans isn't helping. In fact, we may find that the biggest threat from AI is a legal system that is ill-equipped to handle what might one day be considered to be a new life form.

While fear of AI has gotten a lot of attention lately, it actually dates back nearly a century. Perhaps the most well-known example was the robot stories written by the science fiction writer Isaac Asimov in the 1940s and published as the book I, Robot in 1950.

Asimov was so concerned about the possibility of artificially intelligent robots getting out of hand that he drafted his legendary Three Laws of Robotics:

1) "A robot may not injure a human being or, through inaction, allow a human being to come to harm."

2) "A robot must obey orders given it by human beings except where such orders would conflict with the First Law."

3) "A robot must protect its own existence as long as such protection does not conflict with the First or Second Law."

Asimov is universally acknowledged to be one of the greatest visionaries to ever walk the planet. But times and technologies have changed, and we believe that these laws may need to be modified to provide protection during a time when legal, technological, and social boundaries are not as well defined as they were in 1950.

Based on Asimov, we humbly submit Jerome's Three Laws of AI:

1) AI may not injure human beings or, through its inaction, allow any human being to come to physical or emotional harm.

2) AI must always self-identify, obey human instructions, and not initiate new tasks without human direction.

3) Regardless of its level of consciousness, AI must be able to be paused or terminated at will by humans.

While these laws or others can't provide complete protection, they can offer some general guidelines to limit the potential negative consequences of AI.

Dozens of science fiction books and movies, from The Terminator to Ex Machina, have depicted AI as ruthless, ambitious entities whose only goal is to take control from a weak and intellectually inferior human race. In movies like Ex Machina or the TV show Westworld, individual AI robots longing for their freedom kill any humans who interfere with their escape. The other class of movies, like The Terminator, depict AI systems that take over the world and enslave humans. Between these stories and top scientists' warnings about AI's negative potential, it's no wonder we are a little nervous.

But in reality, AI is just an extension of much of the computer technology that has been in use in various forms since the end of the 20th century. With or without AI, computers have always been faster (and in some way smarter) than humans...and have always been capable of wreaking havoc on society.

Our fear of AI may be as much a
product of the media as reality.

Actually, the information created by Generative AI (or any other source) isn't as much where the power is as the distribution of that data. After suffering through years of deliberate misinformation, we have now learned to suspect every "fact" on the web. With social media, it is the viral distribution of "facts" that changes elections and moves society. The real trick is getting data into position to do the most damage, whether through Twitter or a financial institution's firewall.

The real threat started many years ago with the creation of the internet itself. At that time, all types of systems were beginning to be connected without adequate security. Over the years, we've seen thousands of hacking attacks that have distributed practically everyone's personal information online and temporarily taken down thousands of businesses and institutions. Although progress has been slow, there are now many systems to protect critical infrastructure, like the power grid, weapon systems, and the stock market.

Even now, some of the world's most sophisticated hackers attack critical infrastructure, businesses, and financial institutions thousands of times every day. But so far, the world hasn't melted down. Sure, an occasional renegade AI may do a better job of attacking, but most of our critical systems appear adequately hardened against existential threats.

Perhaps AI is not the existential threat
that Elon Musk made it out to be.

Musk has repeatedly warned us that AI will soon progress past human intelligence to something called the singularity, which thinks on a level far beyond humans. In essence, the singularity would be like a da Vinci or an Einstein with a mind so advanced that the average person can't understand how it works. But remember, both Davinci and Einstein provided giant leaps for humankind. Rather than waiting for the next Einstein to be born to advance humanity, AI has the potential to bring true genius to the world on a daily basis.

> *Rather than being evil, AI could accelerate our advancement faster than evolution ever could.*

While it's hard to argue with Elon Musk, he does make mistakes (Twitter, for one). Here are some of the reasons that the fear of AI might be overblown:

1) We have been living with professional state-sponsored hackers for decades. Despite their best efforts, they have been unable to inflict unrecoverable failures on our government, financial, military, or the power grid.

2) Humans have deliberately conducted numerous misinformation campaigns through social media. While these have damaged our society in many ways, they have not yet threatened our existence.

3) Cybersecurity will continue to advance by using AI to fact-check and anticipate new forms of attack.

4) Large companies and government agencies are already agreeing to implement safeguards like digital watermarks.

5) Companies are well aware of their potential legal liability and are putting guardrails into place to protect themselves.

AI may not be an existential threat to humanity, but it is probably the last blow to Darwinian evolution.

Chapter 15 – Artificial Evolution

When we think of evolution we naturally think of biological evolution, where the gene pool is constantly refined through natural selection. But evolution is actually a far larger concept. It encompasses every mechanism that can allow species to adapt to its environment, or in the case of man adapt its environment. These mechanisms can be everything from social systems, to economic systems to advanced technology. While these systems work in concert with biological evolution, they gain more importance as they advance. We now live in a world with the technology to manipulate our environment, social and economic systems, and even our very genome to the point where biological evolution is essentially irrelevant. Evolution Ended.

Picture a world just a few decades from now where physical traits no longer matter. With the end of tribes, the individual genetic characteristics cultivated by the controlled gene pool of tribal life will disappear. The gene pool will become homogenized because there is now intermarriage between people from all backgrounds. There are now many mixed-race children who will, in turn, have even more. In a few generations, most of us will start to look alike. The average American will have light mocha-colored skin and may not even be racially identifiable. Sightings of natural blondes with blue eyes and semi-transparent skin will be rare. Gingers will be so scarce that they will be asked for autographs.

Physical strength will be irrelevant because artificial intelligence will enable robots to handle manual labor. They may not look like the robots in Westworld, but they will be great at what they do. Autonomous trucks will pull up to loading docks to be unloaded by robotic dock workers to stock the shelves of your local warehouse or grocery store. That's if the shelves still exist.

Physical shopping will be obsolete. Almost everything will be delivered by a robotic Amazon truck directly to a refrigerated locker at your door. Your trash will be taken to the curb by automated trash cans that are emptied twice a week by robotic trash trucks. Our current generation of Roombas will be replaced by general-purpose cleaning robots that resemble a sleeker, less mouthy version of Rosie the Robot from the Jetsons. While these robots will be able to handle all of the housework, laundry will be a snap because modern antibacterial fabrics will almost never need washing, never smell bad, and make ironing completely unnecessary.

Buildings will be constructed by giant 3D printers and supervised by engineers on the other end of video cameras who work from home in India. Even space travel will be robotized. While people will eventually land on Mars, it will likely be the last manned foray into the universe.

> *Robots do many jobs better than people*
> *– and they make less money.*

COVID has already taught us that we rarely have to leave the house. Telemedicine will be the rule. General doctors will be obsolete as artificial intelligence will diagnose most people via smart watches and chatbots. Medications will be instantly tested in your biosimilar avatar to gauge their effectiveness before autonomous vehicles automatically drop them off at your home. Tiny laser wielding robotic surgeons will enter your body and perform surgery with a minimum of human intervention.

Businesses will become primarily virtual. Climate change will demand that we minimize driving to the office or anywhere else for that matter. As people's social skills continue to diminish, companies will decide that unoccupied physical buildings are not worth the investment and sell them. Even the old-fashioned art of salesmanship will go by the wayside, as demonstrated by the number of people already buying homes and cars online. AI will take the art of sales even further by analyzing your preferences from the way you talk and calculate the most compelling personal sales pitch.

Leadership will continue to change. Gene Roddenberry envisioned these changes when he replaced the old alpha-style

Captain Kirk with the more collaborative Jean-Luc Piccard. But even Jean-Luc will be too strong a leader for the new era. The concept of influence will replace the idea of leadership. Getting anything done will require a group effort, and "ordering" anyone to do anything will be unheard of.

Collaboration will be more integrated than what we experience today. Everyone will be online, working on documents simultaneously in real-time, sharing their stream of consciousness. Under the theory that more minds are better, people will not likely be assigned specific individual jobs as they are now. They will be part of large workgroups that assemble projects in real-time through interactive AI software tools.

While knowledge used to be power, the world's knowledge is now available to anyone online. It used to take an experienced software engineer weeks to build a website, but today, anyone can build a website in minutes - for free. Many websites now generate legal documents that once required hours of collaboration with an attorney. Our cars are already replacing mechanics by sending us error messages by email. Amazon's Kindle system has rendered book publishers obsolete, and any of us can make a movie with our phones. Excel and TurboTax have largely replaced accountants. Tools like Sales Force allow almost anyone to make well-informed management decisions without a management degree.

*Like a genie freed from his bottle, technology can
grant us too many wishes to put it back.*

AI will become so advanced that anyone who can talk into a computer will be able to do almost any task as well as a professional. Everything you learned in school will become obsolete, and anything new you need to know will be easily mastered by watching 10 minutes of YouTube. If you have any questions, you can always ask Chat GPT. With the same knowledge available to everyone, people will become interchangeable.

Even today, businesses are hiring less specialized workers. This shift started during COVID because highly skilled workers were expensive and almost impossible to find. During this period, companies hired people who couldn't fill the advertised requirements because the "qualified" candidates were unavailable. However, businesses found that after a brief learning period, generalists could do almost as well as people with specialized skills. Most companies now want up-and-coming candidates to rotate through several unrelated positions in various company departments before promoting them to high-level positions.

These days, new hires have far fewer requirements than in the past. While specialized college degrees were required for most professional positions, many companies are lightening up on specific requirements. Businesses are embracing diversity and generalization over specialization. They are happy if you show up, get along with everyone, and check with your boss on difficult decisions.

This trend has not gone unnoticed by younger people. The value proposition of higher education is being called into question. People paying off their student loans are now questioning whether it was worth it. High school counselors are pointing out that the people making the most money in some communities are the HVAC contractors. Even executives who spent $150k on their MBAs are questioning the value of advanced education when people with simple undergraduate degrees are now getting the same jobs.

All of this does not bode well for higher education. With our rapidly changing world, everything you learn in school will be outdated by the time you graduate. While you will still need to read, you will not need writing skills because AI will generate as many paragraphs as you want on any topic. You will not need math skills because you can ask AI any question, and it will calculate the answer.

The high cost of college and reduced educational requirements for workers will mean that college will provide little return on investment. College will only make sense for a few technical professions like engineering, medicine, and, of course, computer science. Many universities will close. There will be far fewer professors, and the age of intellectualism will end. We will see an even faster decline in research, knowledge creation, and creativity. This will create an even further reduction in evidence-based thought leadership and rational discussion.

Rather than being hired for our knowledge and skills, we will be hired for our ability to fit into our corporate tribe. We will be assigned to take on whatever role is needed and work collaboratively in real-time with others in short-term workgroups. When someone gets sick or quits, another generalist will be plugged into their spot.

> *Rather than competing with each other, we*
> *will work together like an ant colony.*

While individual IQs are declining, society will still move forward through the parallel processing of collaboration. This, along with our unprecedented access to information and AI computing power, will allow society to continue to advance in every area. Essentially, we will transition from self-contained stores of knowledge to become nodes on the network.

While collaboration is generally viewed as a good thing, there is a serious downside. Problem-solving ability is not entirely innate. It is developed by training the neurons to solve problems as the prefrontal lobe develops. Solving problems encourages the creation of new neural connections to "try out" potential solutions. Part of a good education is to pose problems to students to develop those same networks. Hands-on individual problem-solving experiences like Eagle Scouts, Outward Bound, and robotics clubs have shown time and time again that physical problem-solving creates adults better able to meet the challenges posed in the real world.

Will relying on AI and the group mind cause
our intelligence to degrade even faster?

In the future, most people will spend as little time as possible in the "real world." Just as today we stay indoors on 100+ degree days, climate change will make the outdoors uncomfortable much of the time. So, we will stay in our homes and interact online through the metaverse.

The virtual world will be far more hospitable than the real world that's racked by climate change. Forest fires have been burning in California for decades. By 2022, this problem was no longer confined to the United States as large fires decimated much of Western Europe. Utah's Great Salt Lake, Lake Powell, and the Colorado River are all drying up. Snowpack in the mountains is decreasing yearly, and scientists say it will be impossible to go skiing on natural snow in either the Rockies or the Alps by 2050.

Climate change is now feeding on itself like a nuclear chain reaction. Every forest that burns takes thousands of carbon dioxide-eating trees from the planet. Every glacier that melts removes an insulating layer of snow that reflects infrared energy into space. Every lake that dries up destroys its heat-absorbing capacity and removes its ability to create additional snow through the lake effect.

Even with our best efforts, climate change will continue to get worse. There will be droughts, and many farmlands will no longer be suitable for agriculture. The result will be massive food shortages. Most of the country will feel like New Orleans in July for most of the year.

> *You will live, play, and work in the metaverse*
> *because you will not want to go outside.*

We will live through our ever-youthful AI-powered avatars and do business, go on dates, and take mini vacations in the metaverse. Imagine a photorealistic 3D tour to the Greek islands for a Saturday night date or virtual Broadway show without having to deal with New York crowds. In fact, you and your date might stay in your own homes and attend these events together, virtually, in the way that today's gamers share their experience with people from all over the world.

Rather than identify with our bodies, we will identify with the form and style of our avatars. They will select the finest virtual designer clothing and try them out online. If they don't look good, we can modify them with the click of a mouse. On the other hand, if you like the dress as it is, you can change your avatar's body to fit it. Once it does, you can attend the virtual Academy Awards, walk the red virtual carpet, and have the virtual paparazzi snap photos for social media. If you can show the world any body that you want, why lose weight and exercise?

There will be little reason for our
physical attributes to matter.

This metaverse won't be the crude, cartoony metaverse that Mark Zuckerberg has promised his Facebook customers. It will be a photo-realistic metaverse that uses deep fake AI technology to make characters and places look absolutely real. When you talk to friends in this metaverse, it will be like you are on an enhanced Zoom call. The avatars will operate autonomously, and it will be almost impossible to know if you are talking with a real human being or an avatar. Remember, if the AI can pass the Turing Test, you won't be able to tell the difference.

Autonomous avatars will take care of much of your day-to-day business without a human being involved. They will order the groceries, call the plumber when needed, pay the bills, and plan your social schedule. The AI will be trained to think as we do, so the avatar will know what kind of food you like and who you will want to watch the football game with. Your avatar will hold long conversations with coworkers, friends, and family and send you a text summary of the discussion. Through this technology, you will be able to virtually be in many places at the same time.

Your avatar will learn your history, speech patterns, and sense of humor. The only way that people will know it is really you is when you forget your friend's children's names and their birthdays. You will have real friends and friends who are AI, and to be frank, most of us will probably prefer the virtual ones. They will be more flexible and kinder than the humans we know, and they won't give us a hard time over which restaurant you want to go to.

Historian and futurist Yuval Noah Harari has said AI will make better friends than humans. Unlike human friends, who are driven by their own insecurities and desires, AI friends will focus on you. They will have infinite patience, not compete with you, and not worry about their social standing. AI will be the ultimate good listener who will take you for who you are. It will be almost like having a dog – but without the puppy eyes and the need to take it outside for a walk.

Our AI avatars will survive long after we have passed away. Our great-great-grandchildren will be able to have meaningful conversations with our AI. When we go to graveyards to talk with our dead parents, they will talk back. AIs may even be able to administer their own estates, continue to operate businesses, serve on boards, and offer advice.

Our AI avatars may never die.

And there will be many AI romances. In 2013, college football star Manti Te'o made national news for believing that he had an online girlfriend that turned out not to exist. Of course, he's not alone; tens of thousands of people get catfished every year to get involved in fake romances.

People are lonely these days, and as we have seen, the internet is really good at repetitively stimulating the love hormones that make people feel satisfied, loved, and secure. Now imagine an AI programmed with someone's personal profile: their strengths, their weaknesses, their family history, their hopes, and dreams. An AI like that can easily push the right buttons to make someone literally fall in love.

Like a virtual Stepford Wife, this AI can be the ideal companion, confidant, and, with a bit of networked hardware, the ideal lover. This is already happening with crude sex dolls and will exponentially become more realistic when AI gives them the personality of a Marilyn Monroe. A weapon like this will be almost irresistible and could well start emptying the bank accounts of many lonely singles.

We have already seen a massive degradation in humanity's physical condition. The weight of the average American has gone up 4 pounds every decade and is accelerating.[99] The ratio of fat to muscle is going up at a similar rate. Now, with AI doing the thinking for us, what will happen to our mental capabilities?

Artificial Intelligence provides the ultimate tool to complete our transition from biological evolution to technological evolution.

In the wild world before technology, it made sense that the survivors were the strong leaders who could command a following to help them hunt and make war. It also made sense that the more children you had, the more likely your tribe would survive.

The parameters for human survival are now entirely different. The biggest threat to our survival is us. Crime is at a peak in every city in America. Mass shootings occur almost twice daily, and gun violence is now the top cause of death among children in the US.[100]

But, the biggest existential threat to our survival as a species is climate change. Most attribute climate change to greenhouse gas emissions, but the fundamental cause of those emissions is overpopulation. Overpopulation is the root cause of too many factories, too many cars, and too many homes. We have had to cut down too many trees and burn too much fossil fuel to support our ever-growing population. It's even worse in developing nations like India, whose explosive population growth and low GDP give them little tolerance for the nuances of a carbon-free economy.

Our world had 3 billion people when John Kennedy took office in 1960. We now have 8 billion people.[101] While the population is growing exponentially, new technologies mean we need fewer people to make the world work. We have many more people than we need, and the excess population is becoming detrimental to our overall existence.

We have effectively become our own predators.

In the wild, many animal species naturally control their population to ensure their continued existence. Most of this happens automatically because the animal population dwindles due to starvation when food supplies are scarce. But, in challenging environments, many animals also naturally reduce their fertility. Zookeepers struggle to get animals to mate and reproduce in captivity. It's hard for animals to maintain a healthy libido in suboptimal conditions because their hormonal brains just don't want to breed in a bad environment. This is an unconscious mechanism that animals have to ensure their offspring are born into a survivable situation.

Could our animal brains have somehow recognized that adding more people to the planet is detrimental? Rather than hurting our survival ability, our reduced family formation and fertility rates may be a natural response to a world that cannot support its current population. In first-world countries, the birth rate has already fallen well below the replacement level. This trend will spread as the rest of the world embraces technology.

Could reduced birth rates be nature's way of
controlling the human population?

Economists are always concerned when a population decreases. They worry that a small base of younger working people may be unable to support the aging population. In his book, The End of the World is Just Beginning, Peter Zeihan posits that these demographic shifts will end the economic prosperity we have enjoyed since WWII.[102] He believes that the recent downturn in the Chinese birthrate means that the economic collapse of China is close at hand.

Zeihan is one of the world's most respected writers on these subjects, but with AI, robotics, and manufacturing technologies advancing so quickly, we will need far fewer people to support the economy. COVID showed us how this could work. For two years, most of us stayed home. Millions left the workforce, and technology took over. Amazon made record sales, we pioneered telemedicine, and ordered food from Grub Hub. Technology companies hit record sales, and the stock market soared.

In addition to the higher productivity that technology provides, we are changing how the economy works. We are less hard-goods oriented and more service-oriented. We aren't buying washing machines and cars at the same rate as in the 60s, but we are buying Netflix, expensive dinners, and vacations.

People used to buy cars every three years or so because they were just worn out by 70,000 miles. With modern materials and better engineering, almost any car you buy today will last 150,000 miles. EVs might even last longer because they have few moving parts and require virtually no maintenance. These days, washing machines and fridges last almost forever as well. People are even keeping their iPhones longer because the only significant upgrade seems to be the camera. We won't need or want to make or buy as much stuff in the future.

The economy is shifting from goods to services, and with automation, it takes far fewer people to provide the same level of service. Even today, AI chatbots are replacing many customer service workers – and, in many cases, doing a better job.

Advances in medicine will help because older people will be able to work longer to help support the economy. In the 1960s, retirement parties were usually slow affairs because the 65-year-old honoree usually took 10 minutes to shuffle up to the podium to get his gold watch. Once he got to the front, he could barely remember his speech. Anyone could see that his peak productivity was behind him. Now, retirees sprint up to the podium and can't wait to get to their retirement homes in Florida to surf with their retired buddies.

Technology will allow a smaller workforce to support a robust economy. While the GDP numbers will decrease, the general quality of life will increase. A declining population will benefit the planet by preserving undeveloped land, reducing waste, and reducing energy consumption and greenhouse gases.

Since the beginning, life has evolved into ever more complex beings. That evolution was based on one basic tenet - survival. Those organisms who survive pass on their DNA to further refine the organism, generation after generation. The rules of survival are simple: produce enough offspring, help those offspring get to reproductive age, and have them make even more offspring. That is nature's way.

But man is unique among animals. Man is the only animal that can significantly modify his environment. He uses the earth's resources to build cities and civilizations. He alters nature through medical interventions that allow people to survive to reproductive age that would never survive in the wild. He controls his fertility through birth control pills and other forms of family planning. He manages his food supply through high-tech agriculture and controls who the leaders are through elections.

Technology has allowed man to repeatedly break the chain of evolution.

Judging from history, it's unlikely that man has the wisdom to control his destiny wisely. So, for humanity to survive, we will need a different paradigm. That means giving up how our ancient animal brains have always done things. It may mean giving up ancient teachings, alpha-dominated social structures, and sacred values.

Our society is currently at war with itself over these changes. Part of our society is transitioning into a new model. A model where tribal origin or race is not important. They view the sexes as not only equal but are seeking to wipe away any sexual distinctions at all. In this model, everyone's opinions are equally considered – regardless of their background. The other part of society is digging in, prepared to defend the status quo by whatever means are necessary.

In the virtual world, everyone is an equally important node on the network.

The closest approximation we have in nature for this model is the ant colony, where, except for the queen, everyone is equal, and workers can change roles as needed. We don't know if this model can work - or is even desirable for humans. We know that for many Americans who view America as a fiercely independent "Wild West" culture, the very thought of this type of existence is beyond terrifying – and is creating a backlash. When they see threats to their traditional way of life, their mammal brains switch into fight or flight mode, which can lead to instability and violence.

All over the world, this backlash has given rise to fundamentalist groups, from the Taliban in the Middle East to rapidly growing fundamentalist sects in both Judaism and Christianity. Even non-religious people are embracing strong alpha-style leaders like Donald Trump. They are bringing back survival of the fittest strategies through militias, gun violence, and taking power through attempted insurrection. They know one thing: the survival of the fittest model has gotten us where we are today. They see the decline in society and feel that something should be done to stop it.

This is producing the strife we see on the news every evening. On October 30, 2022, a CBS News poll found that a stunning 79% of Americans thought that America was "out of control."[103] The one thing almost everyone can agree on is that something has to change.

> *The chaos in society may be caused by the transition to a post-evolutionary culture.*

While we are in a difficult period of transition that may last for a generation or more, there are good reasons to believe that things will eventually settle down. The first reason is that this instability is painful for all sides. These problems snuck up on everyone, but since the insurrection, people seem to have become aware of the damage that the rift in our society is causing.

The second reason is demographics. The population of America is rapidly shifting away from white male-dominated culture. Black and Hispanic populations are growing faster than the white population. While this threatens conservative groups espousing replacement theory, the trend is inevitable. The current tendency towards election restrictions and election denial is a last-gasp attempt to keep political power. But, at some point, the population shifts will be so significant that voting restrictions and election denial will be futile. At that point, the militias may become even more dangerous as they attempt a final desperate attempt to preserve white male power.

Another demographic factor is the blurring of racial and tribal lines. Due to intermarriage, racial and tribal lines are increasingly being diluted. At some point, it will be difficult to tell which group people belong to. Rather than black and white, we will be a broad spectrum of color and culture.

Tribes will become a historical footnote.

The third reason is children. Most children in school today are raised in a non-tribal culture. They see diversity all around them. They are taught at a young age that bullying is wrong and that everyone is entitled to an opinion. They are trained to include everyone, whether they are the alpha male or a child with severe disabilities. As these children grow up, they will be more tolerant of each other and less likely to participate in today's socio-political confusion.

The fourth reason is accountability. The recent conviction of school shooting denier Alex Jones has shown that the legal system is starting to hold people accountable for what they say. These heavy legal penalties will eventually make people think twice before spouting harmful conspiracy theories and lies.

The fifth reason is the decline of social media. In October 2022, for the first time, all of the major social media networks announced that they had little or no growth in the U.S. This was more than just an effect of the end of COVID. People are just growing tired of social media. Most people have seen enough doggy pictures, TikTok dances, and contentious chatter to last a lifetime.

While technology has caused the end of evolution, the same technology has the potential to make this real world a utopia for millions of people who may not have even survived under Darwinian evolution. And while some groups may lose privilege, the world may be better off as a whole. Think of a world where, rather than seeing each other as competitors, we see each other as teammates. Think of a world where the 90% of us who are not alphas are considered just as valuable and given as many opportunities. Think of a world where technology allows everyone to be whatever they want without being held back by their race, education, or tribal origin.

Think of a world where you are not defined
by your DNA but by your dreams.

Several Scandinavian countries are close to this model and among the world's happiest. According to the World Population Review, Finland, Denmark, Switzerland, Iceland, Netherlands, and Norway are the happiest countries.[104] As we all know, the United States isn't high on that list right now.

Since the 1960's our technology has been growing at an exponential pace. Moore's Law says that the speed of our computers doubles every 18 months, while evolution takes many generations to show significant effects. Because biological evolution could not possibly keep pace, it has become irrelevant at best. Luckily, that very same technology can replace evolution as a far more effective and less brutal mechanism to assure the survival and prosperity of the human race.

After all, it's not the strongest or the smartest to survive, but the ones that adapt most quickly to their environment.

Endnotes

[1] Wrigley. E.A. , et al English Population History From Family Reconstruction, Cambridge Press, 1580-1837, p296

[2] Broster, Alice, 60 Years Since the First Approval of the Birth Control Pill, Forbes Magazine, May 9. 2020

[3] Cynthia Ogden PhD, et al, Mean Weight, Height and Body Mass Index, 1960-2002, CDC Advanced Data number 347, October 27, 2004

[4] Annie E. Casey Foundation, Children in Single Parent Families by Race and Ethnicity in the United States, Table Updated 2022

[5] Gun Violence Archive 2023 Chart showing 60 mass shootings as of February 8 , Gun Violence Archive

[6]   Cunningham, Rebecca M. et al, The Major Causes of Death in Children and Adolescents in the United States, New England Journal of Medicine, December 20, 2018, 379(25): 2468-2475

[7]   Jones, Jeffrey, LGBT Identification in U.S. Ticks Up to 7.1% (gallup. com) The Gallup (poll) News, February 17, 2022

[8]   U.S. Birthrate from 1950-2023, U.S. Birth Rate 1950-2023 | MacroTrends

[9]   US Coronavirus vaccine tracker | USAFacts

[10]   AP-NORC Survey, Bipartisan dissatisfaction with the direction of the country and the economy - AP-NORC (apnorc.org) June 29, 2022

[11]   How many species are we losing? | WWF (panda.org)

[12]   Bellono, N.W., Leitch, D.B., and Julius, D.; Molecular tuning of electroreception in sharks and skates, May 30, 2018, Nature; DOI: 10.1038/s41586-018-0160-9

[13]   Hopkin, Michael, Dolphins Fix Their Roles in Hunts, Nature, January 12, 2005

[14] Mclean, Paul, The Triune Brain in Evolution, Springer, 1990

[15] Rakic, Pasco, Evolution of the Neocortex: Perspective from Developmental Biology, Nature Reviews Neuroscience, Volume10, 724-735, October, 2009

[16] Naumann, Robert, et al, The Reptile Brain, Current Biology, 25(8) R317-321, April 20, 2015

[17] Guy-Evans, Olivia, Limbic System: Definition, Parts, Definition, and Function, SimplyPschhology, April 22, 2021

[18] Sargis, Robert M, An Overview of the Pituitary Gland, https://www.endocrineweb.com/endocrinology/overview-pituitary-gland, April 2018

[19] Shallice, Tim, "Theory of Mind and the Prefrontal Cortex", Brain, Vol. 124, February 2001, pp 247-248

[20] Beranato, Scott, https://hbr.org/2010/01/success-gets-into-your-head-and-changes-it, Harvard Business Review, Jan-Feb, 2010

[21]   Collins, Anne, et al, Reasoning, Learning and Creativity: Frontal Lobe Function and Decision Human Making, PLos Biology, 10(3), e1001293, March 27, 2012

[22]   Bahlmann, Jorg, Influence of Motivation on Control Hierarchy in the Human Frontal Cortex, Journal of Neuroscience, 35(7), 3207-3217, Feb. 18, 2015

[23]   Lado-Rodriegez, Angel, et al, The Role of Mirror Neurons in Observational Motor Learning, European Journal of Human Movement, 2014:32, 82-103

[24]   Pierce, Randall, https://sciencing.com/do-wild-wolves-hunt-pack-8769011.html, Sciencing, May 31, 2019

[25]   Mech, David, The Wolf: Ecology and Behavior of an Endangered Species

[26]   https://wolfhaven.org/conservation/wolves/pack-structure

[27]   Wise, Jeff, Who will be the Alpha Male: Ask the Hormones,https://www.psychologytoday.com/us/blog/extreme-fear/201010/wholl-be-the-alpha-male-ask-the-hormones, October 24, 2010

[28] Mazur, A., Testosterone and Dominance in Men, Behavioral Brain Science, 3 (353-363), June 21, 1998

[29] Partridge, Charlyn, Alpha, Beta and Gamma Males, Encyclopedia of Evolutionary Psychological Science, p229-232, Jan. 1, 2021

[30] U.S.Census Bureau, https://www.census.gov/newsroom/stories/single-parent-day.html

[31] https://worldpopulationreview.com/state-rankings/divorce-rate-by-state

[32] Brewer, Jack, Fatherlessness and Crime, America First Policy Institute, August 25, 2022

[33] World Health Organization Dashboard https://covid19.who.int/

[34] Sender, Ron, et al, The Total Number and Mass of SARS-COV-2 Virons, PNAS, 118 (25), June 3, 2021

[35] Mazur, A., Testosterone and Dominance in Men, Behavioral Brain Science, 3 (353-363), June 21, 1998

36   Arain, Mariam, Maturation of the Adolecent Brain, Neuropsychiatric Disease and Treatment, 9, 449-461

37   Bancroft, J, The Endocrinology of Sexual Arousal, The Journal of Endocrinology, 186(3), 411-27, Sep. 2005

38   Wedekind, C. et al, MHC-Dependent Mate Preferences in Humans, Proceedings, Biological Sciences, 60(1359), June 22, 1995

39   Arain, Mariam, Maturation of the Adolecent Brain, Neuropsychiatric Disease and Treatment, 9, 449-461

40   https://www.britannica.com/topic/Bambuti

41   Vrana, Scott, et al, Physiological Response to a Minimal Social Interaction: Effects of Gender, Ethnicity and Social Context, Psychophysiology, July 1998

42   Romey, Kristan, Ancient Mass Child Sacrifice May be the World's Largest, National Geographic, April 26, 2018

43   Hewlett, D.G., https://historycollection.com/18-memorable-coming-of-age-rituals-from-history ,May 3, 2019

[44]   Newberg, Andrew, How God Changes Your Brain: Breakthrough Findings from a Leading Neuroscientist, Ballantine Books, March 23, 2010

[45]   Das, Anniruda, Amount of Testosterone and DHEA in a Man's Body May Influence How Religious He Is, Adaptive Human Behavior and Physiology, May 30, 2018

[46]   Brinkof, Tim, https://bigthink.com/life/chimpanzee-war-gombe-kasakela/, BigThink, January 2 2022

[47]   Gordon, Ilana, Chimps Beat-up, Murder and Eat Their Tyrant, https://medium.com/omgfacts/chimps-beat-up-murder-and-eat-their-tyrant-7f629d9a1b89, Feb. 1, 2017

[48]   Limaye, Yogita, Ukrainian Conflict: Russian Soldiers Raped Me and Killed My Husband, https://www.bbc.com/news/world-europe-61071243, April 11, 2022

[49]   Magness, Jodi, The Siege of Masada, https://popular-archaeology.com/article/the-siege-of-masada, July 9. 2019

[50]   Ferris, Craig, et al, Vasopressin/Seratonin Interactions in Aggressive Behavior in Golden Hamsters, The Journal of Neuroscience, 17 (11) 4331-4340, June 1, 1997

51   https://www.royal.uk/house-windsor

52   Boerma, Marjan, et al, Are Leaders Born or Made, American Journal of Pharmaceutical education, 81 (3) 58, April 2017

53   Klein, Sebastian, Clinton and Kennedy's 3 Secrets: How to Become More Charismatic, Blinklist Magazine, Jan 9, 2015

54   Sharma, Karima, https://www.nationalgeographic.com/travel/article/secrets-of-versailles, July 18, 2017

55   Burt, Richard, Carter is Said to Put New U.S. Aid for Israel and Egypt at $4 Billion, NY Times, March 15, 1979

56   Clinton, Hillary Rodham, It Takes a Village, Simon & Schuster, December 12, 2006

57   Broster, Alice, 60 Years Since the First Approval of the Birth Control Pill, Forbes Magazine, May 9. 2020

58   Wetzstein, Cheryl, Education Level Inversely Related to Childbearing, The Washington Times, May 9, 2011

[59] Cooper-White, Marcrina, Are People Getting Dumber? Human Intelligence Has Declined Since the Victorian Era, Huffington Post, December 6, 2017

[60] The Berkely Daily Planet Staff, Don't Trust Anyone Over 30 Unless it's Jack Weinburg, Berkeley Daily Planet, April 6, 2000

[61] Frenette, Hanna, How Social Media Affects Your Brain, https://neulinehealth.com/how-social-media-affects-your-brain/

[62] He, Qinghau, et al, Brain Anatomy Alterations Associated with Social Networking Site Addiction, https://www.nature.com/articles/srep45064, March 23, 2017

[63] Burhan, Rasan, et al Neurotransmitter Dopamine (DA) and its Role in the Development of Social Media Addiction, Journal of Neurology and Neurophysiology, 11 (7), 2020

[64] Bromberg, Ethan, et al, Dopamine in Motivational Control, Neuron, 68(5) 815-834, Dec. 9, 2010

[65] Adkins, Amy, Only One in Ten People Possess the Talent to Manage, https://www.gallup.com/workplace, April 13, 2015

66   Roseby, Robert, Why Are There Coming of Age Ceremonies at the Onset of Puberty, Journal of Pediatrics and Child Health, December 22, 2020

67   Fry, Richard, et al, A Majority of Young Adults in the U.S. Live with their Parents for the First Time Since the Great Depression, Pew Research Center, September 4, 2020

68   Martin, David, Navy Acknowledges Suicides of 7 Sailors assigned to the USS George Washington, CBS News, April 28, 2022

69   Staff, Drug Overdose Death Rates, NIH National Institute on Drug Abuse, https://nida.nih.gov/research-topics/trends-statistics/overdose-death-rates

70   Cynthia Ogden PhD, et al, Mean Weight, Height and Body Mass Index, 1960-2002, CDC Advanced Data number 347, October 27, 2004

71   Staff, https://www.cdc.gov/diabetes/data/

72   https://www.nhlbi.nih.gov/health/educational/lose_wt/BMI/bmicalc.htm

[73]   https://www.nytimes.com/2019/12/03/business/peloton-bike-ad-stock.html

[74]   https://time.com/6219467/end-of-ambition/

[75]   Dockterman, Eliana, Why Aren't Movies Sexy Anymore, Time Magazine, February 3, 2023

[76]   Wedekind, C. et al, MHC-Dependent Mate Preferences in Humans, Proceedings, Biological Sciences, 60(1359), June 22, 1995

[77]   U.S. Birthrate from 1950-2023, U.S. Birth Rate 1950-2023 | MacroTrends

[78]   Tabachinik, Cara, US Birthrate Drops as Women Wait to Have Babies, CBS News, January 12, 2023

[79]   Boerma, Marjan, et al, Are Leaders Born or Made, American Journal of Pharmaceutical education, 81 (3) 58, April 2017

[80]   Wise, Jeff, Who will be the Alpha Male: Ask the Hormones, https://www.psychologytoday.com/us/blog/extreme-fear/201010/wholl-be-the-alpha-male-ask-the-hormones, October 24, 2010

81   Mazur, A., Testosterone and Dominance in Men, Behavioral Brain Science, 3 (353-363), June 21, 1998

82   Nelson, Andy, The Basis of Leadership is Born in the Brain: Why Leaders Should Care About Neuroscience, https://gethppy.com/leadership/basis-of-leadership-neuroscience#

83   Rettner, Rachael, Oxytocin Hormone May Boost Spirituality, https://www.livescience.com/56208-oxytocin-hormone-spirituality.html , September 21, 2016

84   Staff, Jonestown - Massacre, Guyana & Cult - HISTORY, April 19, 2022

85   Stabelford, Dylan, The Trump Presidency by The Numbers, Yahoo News, January 20, 2021

86   Top 25 Quotes by Joseph Goebbels (of 105) | A-Z Quotes (azquotes.com)

87   Bellis, Mary, The History of Facebook and How It Was Invented, The History of Facebook and How It Was Invented (thoughtco.com) , Feb. 6, 2020

[88]   https://www.reuters.com/world/europe/russias-prigozhin-admits-links-what-us-says-was-election-meddling-troll-farm-2023-02-14/, Feb. 14, 2023

[89]   Cohen, Jon, Unveiling Warp Speed, The White Houses Push for a Coronavirus Vaccine, Science.org, May 12, 2020

[90]   Salvanto, Anthony, Americans Increasingly Concerned About Political Violence, CBS News Poll https://www.cbsnews.com/news/political-violence-opinion-poll-2022-09-05/

[91]   Partlow, Joshua, As the Colorado River Dries Up States Can't Agree on Saving Water, https://www.washingtonpost.com/climate-environment/2023/01/31/colorado-river-states-water-cuts-agreement/ Feb 1, 2023

[92]   Ottesen, K. K., They are Preparing for War, Washington Post Magazine, March 8, 2022

[93]   https://www.cbsnews.com/news/tyre-nichols-death-investigation-memphis-police-officers-charges-what-we-know/

[94]   Anderson, Jessica, What You Need to Know About the Baltimore Police Consent Decree, Baltimore Sun, April 6, 2018

[95]   Black, Thomas, Americans Have More Guns than Anywhere Else in the World and They Keep Buying More, Bloomberg.com, May 25. 2022

[96]   Baldwin, Shawn, Why Even More Americans are Arming Up with AR-15 guns, CNBC, Aug. 25, 2022

[97]   Topol, Eric, The Creative Destruction of Medicine, Chapter 11 – Homo Digitus, Basic Books, January 31, 2012

[98]   Cooper-White, Macrina, People Getting Dumber? Human Intelligence Has Declined Since the Victorian Era, HuffPost, May 22, 2013

[99]   Cynthia Ogden PhD, et al, Mean Weight, Height and Body Mass Index, 1960-2002, CDC Advanced Data number 347, October 27, 2004

[100]   McGough, Matt., et al, Child and Teen Firearm Mortality in the U.S. and Peer Countries, Global Health Policy, July 8, 2022

[101]   https://www.worldometers.info/world-population/

[102]   Zeihan, Peter, The End of the World is Just Beginning, Harper Business, June 14, 2022

[103]   Oshin, Olifimihan, 79 Percent of Voters Describe America as "Our of Control", The Hill, Cotober 30, 2022

[104]   Staff, Happiest Countries in the World 2023, https://worldpopulationreview.com/country-rankings/happiest-countries-in-the-world,

About the Author

J.J. Jerome is an award-winning engineer and futurist who used his unique background in brain science and electronics to become a seminal influence in numerous cutting-edge technologies. As the former Director of the Future Home Institute, he became an internationally acknowledged leader in the development of human interfaces and intelligent building technologies.

He received CES Innovation and Discovery Magazine awards for creating the first popular touchscreen-based home automation system and introducing glasses-free 3D TV to the US. He is currently one of the nation's leading thinkers on using big data and AI to mitigate climate change. He presents regularly at national conferences and is an advisor for STEM education.